Principles of
Machine Dynamics

Page 47, Figure 2-35:
$\vec{F}_{41}$ should be $\vec{F}_{14}$

Page 48, Figure 2-36, Loop 2:
$\vec{R}_3$ should be $\vec{R}'_3$

Page 77, column 1, line 11:
distance as W_1, Figure 3-1B.

Page 81, column 1, line 30:
$\Sigma\vec{F} = 0;\ W_1\vec{R}_1 + W_2\vec{R}_2 + W_3\vec{R}_3 + W_B\vec{R}_B +$

Page 82, Figure 3-8:

z should be $-z$

$\theta_2 = 210^\circ$ should be $\theta_3 = 210^\circ$

$\theta_3 = 300^\circ$ should be $\theta_2 = 300^\circ$

Page 87, column 1, line 16:
$\theta_3 = 47.3^\circ;\ \theta_3 = -79.5^\circ;\ \theta_4 = 115.7^\circ$ and $\theta_4 = 212^\circ$

Page 88, column 2, line 37:
31 Value of A_{G3} - Accel. of gravity center (for $\alpha_2 = 0$)

Page 90, Steps 101 and 102:
Use P/R Conversion

Page 93, Steps 101 and 102:
Use P/R Conversion

Page 95, Steps 459 and 460:
Use P/R Conversion

Page 31, column 1, lines 3 through 12:

$\therefore \vec{F}''_{34} = 34.49\vec{i} + 19.46\vec{j} = 39.6 \underline{/29.43^\circ}$ lb

Also

$\Sigma\vec{F}_4 = 0$ yields

$\vec{F}''_{34} + (-m_4\vec{A}_{G4}) + \vec{F}''_{14} = 0$

Or

$34.49\vec{i} + 19.46\vec{j} - 66.52\vec{i} + 18.79\vec{j} + \vec{F}''_{14} = 0$

$\vec{F}''_{14} = 32.03\vec{i} - 38.49\vec{j} = 50.07 \underline{/-50.23^\circ}$ lb

$\vec{F}_{32} = \vec{F}'_{32} + \vec{F}''_{32} = -\vec{F}'_{23} - \vec{F}''_{23} = -129.99\vec{i} +$
$25.09\vec{j} - 34.49\vec{i} - 19.64\vec{j} = -164.48\vec{i} +$
$5.45\vec{j} = 164.57 \underline{/178.10^\circ}$ lb

Page 31, column 1, lines 15, 16, and 17:

$\vec{T}_2 = \vec{R}_A \times \vec{F}_{32} = -(-2.6\vec{i} + 1.5\vec{j}) \times$
$(-164.48\vec{i} + 5.45\vec{j})$

$\vec{T}_2 = -232.55\vec{k}$ in.-lb ANS.

Page 31, column 1, line 22:

$25.09\vec{j}) = -65.23\vec{k} + 194.99\vec{k} =$

Page 31, column 2, second line from bottom:

$F_{23} = 131.62$ lb $\underline{/-12.06^\circ}$

Page 32, column 2, lines 8, 9, and 10:

$T''_2 = F_{34}\, b \sin(\theta_2 - \theta_3) = (-40.32)(3) \sin 120 =$
-104.75 in.-lb

Total Torque $T = T' + T'' = -226.11$ in.-lb

Page 33, column 2, line 12:

($m_4 = 2.66$ lb-ft/sec^2 or slugs). Using a $K_s - 1'' = 1'$

Page 36, Figure 2-24:

$I_y = 1.298$ ft^4 should be $I_{\bar{y}} = 1.298$ ft^4

$I_x = 1.576$ ft^4 should be $I_{\bar{x}} = 1.576$ ft^4

Page 37, Figure 2-26:

$I_x = 1.010$ ft^4 should be $I_{\bar{x}} = 1.010$ ft^4

$I_y = 0.367$ ft^4 should be $I_{\bar{y}} = 0.367$ ft^4

Page 7, column 2, line 29:
distance of $h = M/F$. Figures 2-2 and 2-3 show other

Page 7, column 2, second line from bottom:
calculations disregarding the force, P, but considering

Page 11, Figure 2-7:
F_{54y} should be $-F_{54y}$; F_{32y} should be F_{23y}; and F_{32x} should be F_{23x}

Page 15, Figure 2-11:
406.92 should be 139.78; 236.9 should be 82.68

Page 17, column 2, line 2:
mation of moments about G, we find $\vec{d} \times \vec{R} = \overline{I}\vec{\alpha}$.

Page 17, column 2, line 17:
"$\overline{I}\vec{\alpha}$." Notice in Figure 2-12C that the shifing of the

Page 19, column 1, line 24:
in Figure 2-13 with two masses, m_3, m_4, and

Page 24, Figure 2-17, the answer should be:
$\therefore T_2 = F_{32} \cdot h = (2.8^k)(5)$
$T_2 = 14{,}000$ in.-lb ANS.

Page 27, Figure 2-18, the answer should be:
$T_2 = (165)(1.38) = 227.7$ in.-lb ANS.

Page 29, column 1, lines 10 and 11:
$\vec{V}_B = \vec{\omega}_4 \times \vec{d} = -464.72\vec{i} - 228.48\vec{j} =$
$517.85 \underline{/-153.8^{\circ}}$

Page 30, column 1, line 6:
$(86.21\vec{k} \times 1/2\, \vec{r}_4)$

Page 30, column 1, line 14:
$-152.57\vec{i} + 70.82\vec{j} = 168.2 \underline{/155.10^{\circ}}$ lb

Page 30, column 2, line 18:
$-m_4 A_{G4} = -1/12 \left(\frac{0.8}{32.2}\right) (32{,}130\vec{i} - 9{,}075\vec{j}) =$

Page 30, column 2, line 20:
$-I_4 \vec{\alpha}_4 = -(0.01)\ (-8{,}268.4)\ \vec{k} = 82.68\vec{k}$ in.-lb

Dear Reader:

Unfortunately, errata do occur--particularly in a book of the scope and technical detail of Principles of Machine Dynamics. We apologize for this inconvenience and if you have any questions whatsoever, please write directly to me.

Clayton A. Umbach, Jr.
Director, Book Publishing

Principles of
Machine Dynamics

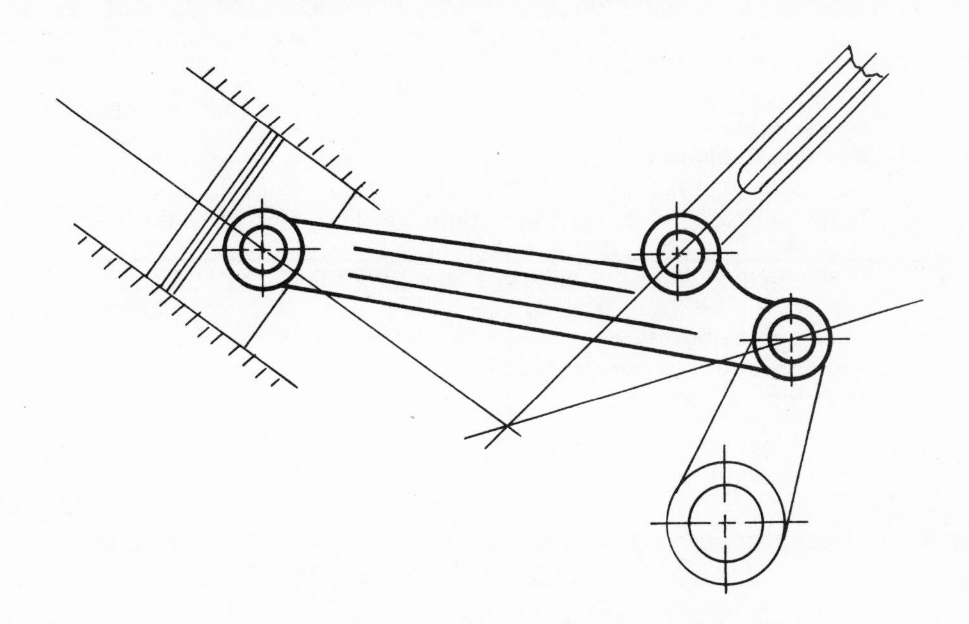

George Raczkowski, Ph.D.
Texas A&M University

Gulf Publishing Company
Book Division
Houston, London, Paris, Tokyo

Principles of
Machine Dynamics

Library of Congress
Catalog Card Number 78-72995
ISBN 0-87201-440-1

Contents

Preface

This workbook is designed for students taking their first course in dynamics of machinery. At Texas A&M University the material in this book is covered in a period of three to four weeks.

The emphasis is on fundamentals and brevity rather than on descriptive detail. Graphical and analytical methods have been used, often repetitiously, to give students a grasp of both.

The major examples and their solutions are oriented toward computer-aided design. They may be considered as a guide to a new dimension in problem solving: the use of the hand-held programmable calculator. A pocket calculator can do many of the engineering calculations needed to complete a dynamic analysis. And it is also important for the solution to be presented in a neat, well organized form. Therefore, neatness, lettering, and line quality were given strict attention. I hope these notes will set an example for students to follow in their professional work.

I shall greatly appreciate comments concerning both errors and suggestions for improvement of this work.

George Raczkowski
College Station, Tx.
February, 1979

1
Glossary of Elementary Kinematics

Kinematic Terminology

$\vec{a}$ or $\vec{A}$	acceleration (vector)
a or A	magnitude of acceleration
a^n or A^n	normal component of acceleration
a^t or A^t	tangential component of acceleration
$\vec{v}$ or $\vec{V}$	velocity (vector)
v or V	magnitude of velocity
$\vec{\alpha}$ (alpha)	angular acceleration (vector)
α (alpha)	magnitude of angular acceleration
θ (theta)	angular position or displacement
ϕ (phi)	angular position or displacement
$\vec{\omega}$ (omega)	angular velocity (vector)
ω (omega)	magnitude of angular velocity

Dynamic Terminology

m	mass; mass per unit length
$I, I_{xx} \ldots$	moment of inertia
$\bar{I}$	centroidal moment of inertia
G	center of gravity; mass center
F or $\vec{F}$	force (magnitude or vector)
M or $\vec{M}$	couple; moment
T or $\vec{T}$	torque (also moment)
W or $\vec{W}$	weight; dead load

Schematic Diagrams

To avoid unnecessary difficulty in showing the true shapes of mechanisms, only the relevant dimensions will be shown on a schematic drawing of the actual machine.

For example, binary, ternary and quaternary links are shown in Figure 1-1. Other symbols represent different types of pairs, such as the revolute pairs and sliding pairs shown in Figure 1-2.

Mobility Criteria for Mechanisms

Grübler's (planar) and Kutzbach's (three-dimensional) equations are used to determine if a mechanism chain is capable of performing any relative motion between links, and the number of degrees of freedom.

Grübler's equation is:

$$F = 3(N - 1) - 2l - h$$

where: F — number of degrees of freedom
N — number of links
l — number of lower pairs
h — number of higher pairs

For example, a four-bar linkage with all lower pairs (Figure 1-3) will have one-degree of freedom according to the formula:

$$F = 3(4 - 1) - 4 \cdot 2 - 0 = 1$$

The cam (Figure 1-4) also has one-degree of freedom:

$$F = 3(3 - 1) - 2 \cdot 2 - 1 = 1$$

Basics of Motion Analysis

Motion may be defined as the changing of position. Velocity is the time rate of change of position. Acceleration is the time rate of change of velocity. Position, velocity, and acceleration are vector quantities.

Two distinct types of motion are observed: translation and rotation. Translational motion may be either recti-

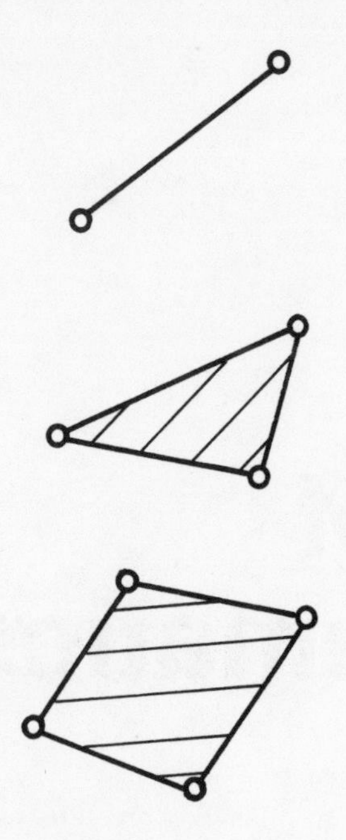

Figure 1-1. Schematic representation of links.

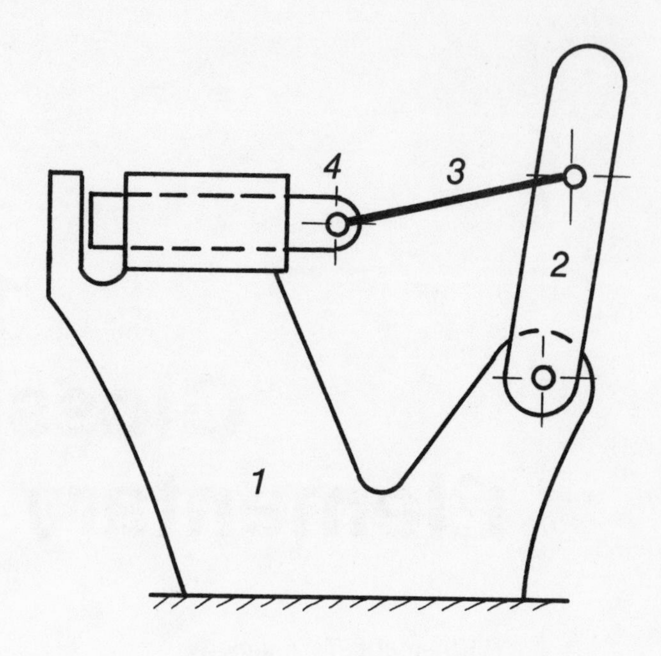

Figure 1-3. One degree of freedom linkage.

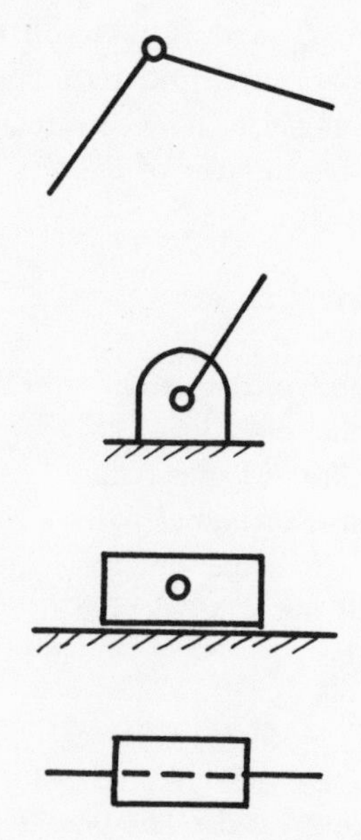

Figure 1-2. Schematic representations of pairs.

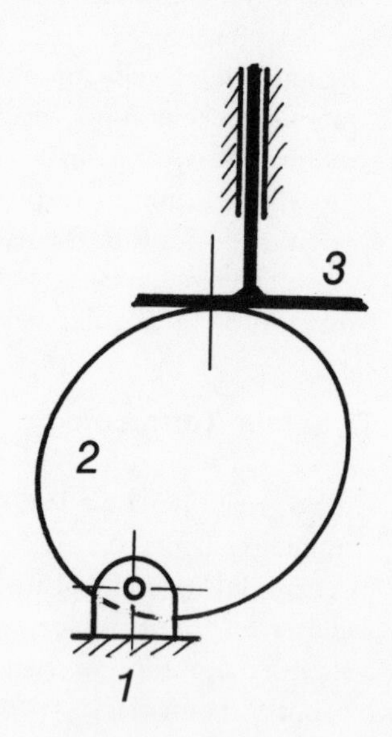

Figure 1-4. One degree of freedom cam mechanism.

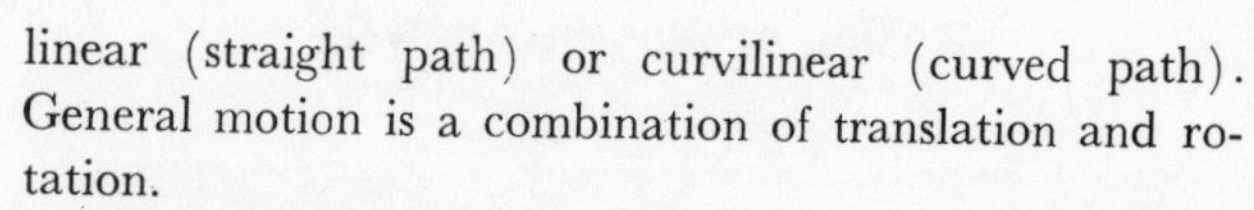

linear (straight path) or curvilinear (curved path). General motion is a combination of translation and rotation.

For example, the different links of the slider crank mechanism will have different motions, i.e., link 1 is at rest, link 2 rotates about point O_2, link 3 has a general planar motion or translation and rotation combined, and finally, link 4 has translation. (See Figure 1-5.)

Relative Motion

The left side of Figure 1-6 shows two positions of a link. Position A′B′ could be reached by:

1. Translation and rotation about A′
2. Rotation about A followed by a translation of the link from A to A′, shown in the figure

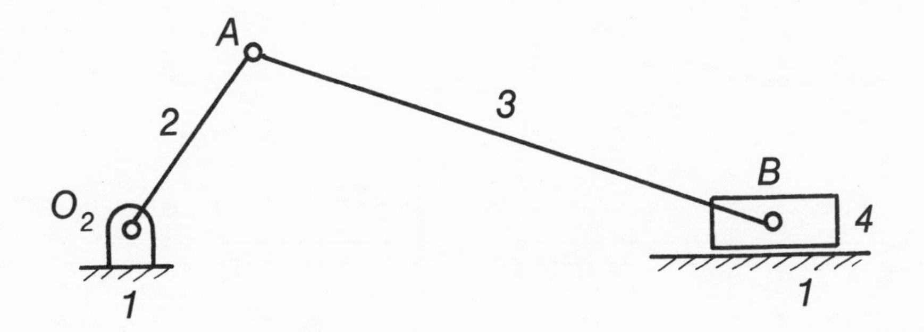

Figure 1-5. Slider crank mechanism.

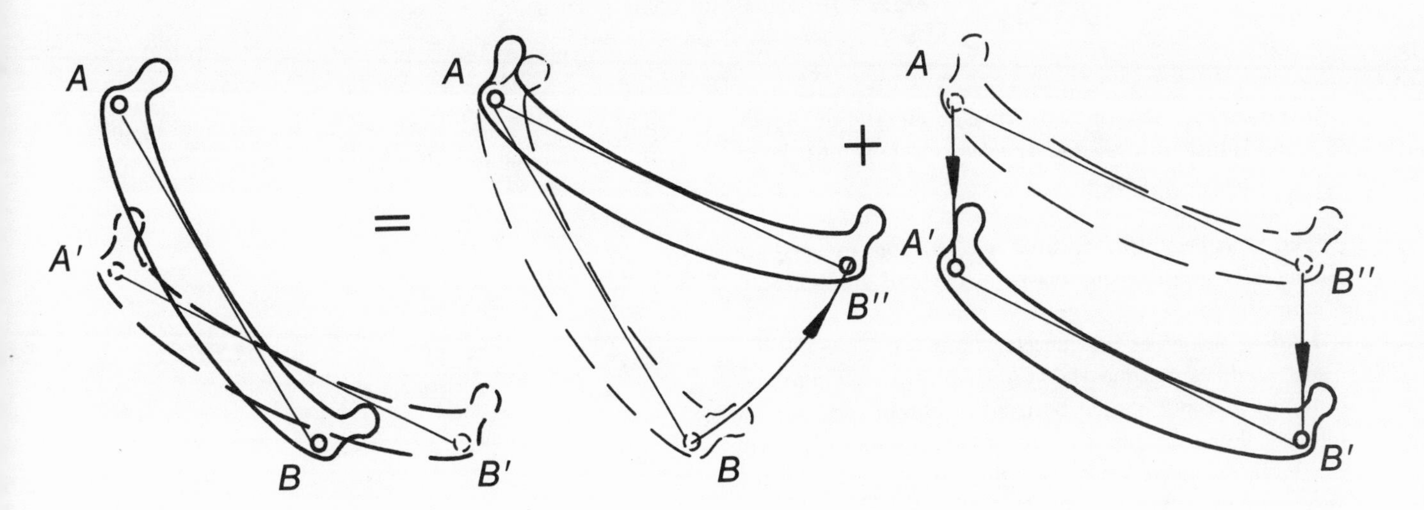

Figure 1-6. Plane motion as a rotation and translation.

The equations describing this motion are obtained from:

$$\vec{S}_A = \vec{S}_B + \vec{S}_{A/B}$$

Differentiating twice, you find

$$\vec{V}_A = \vec{V}_B + \vec{V}_{A/B}$$

$$\vec{A}_A = \vec{A}_B + \vec{A}_{A/B}$$

where the terms with single subscripts (S_A, V_A, A_A, etc.) are called absolute positions, velocities, and accelerations respectively; while the terms with double subscripts are called relative positions, velocities and acceleration components. The equations are extremely useful and true for any two points on a single rigid link. (See Figure 1-7 and solve for $\vec{V}_{B}$.)

Velocity and Acceleration Analysis

Finding velocities and accelerations for different links in mechanisms may be done in two ways:

1. Graphically
2. Analytically

Commonly used methods for finding velocities are by use of:

1. Instant Centers Method (Aronhold-Kennedy theorem)
2. Resolution Method
3. Relative Velocity Method (velocity polygons)
4. Analytical Method (by differentiating the position equations)

Examples of application of the analytical and relative velocity methods are given on pages 4, 29, 44, and 86 of this text.

For determining accelerations of different points and links in mechanisms, the following methods are used:

1. Analytical—differentiating velocity equations as shown on pages 4, 5, 49, and 86.

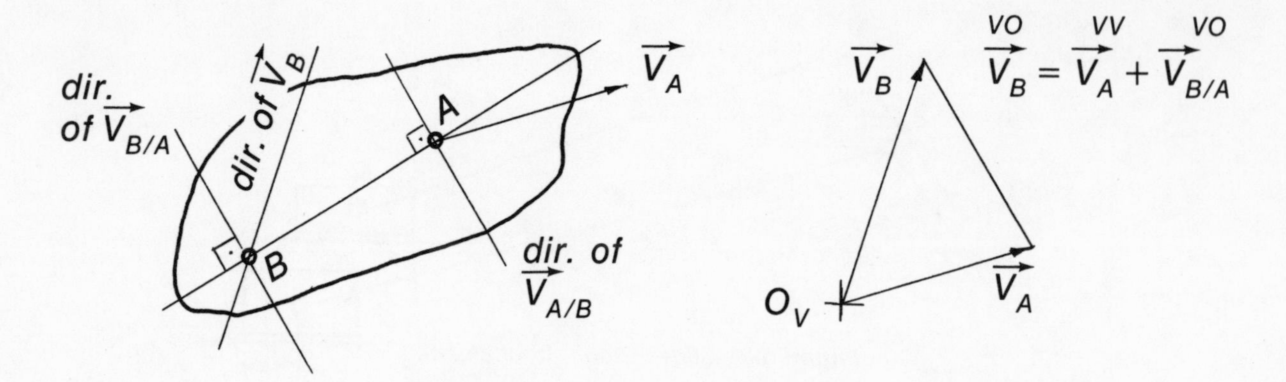

Figure 1-7. *Relative velocity of a link.*

2. Graphical—use of the relative acceleration equation previously discussed or acceleration polygons shown on pages 29 and 45.

The so-called velocity and acceleration images are particularly useful in finding velocities and accelerations of points of interest, e.g., gravity centers.

Finding velocities and accelerations for some mechanisms could sometimes be a difficult task. Therefore, different methods should be tried to obtain the simplest solution. For example, the use of complex numbers will eliminate the need for finding the Coriolis acceleration, and it is particularly suitable for programming.

In order to determine the inertia forces and torques, a correct solution for the linear and angular accelerations is needed.

A well done kinematic analysis is a prerequisite for any dynamic force analysis.

Four-Bar Linkage Example

Another four-bar linkage commonly used is the offset slider-crank mechanism shown in Figure 1-8 and Chapter 2, Review Problem 14.

For the purpose of dynamic analysis study, a purely analytical method will be used for determining the equations for ω_3, α_3, $\dot{x}$, and $\ddot{x}$. Taking y components for the mechanism as shown in Figure 1-8 gives:

$$b \sin \theta_2 = a - c \sin (180 - \theta_3)$$

or

$$c \sin \theta_3 = a - b \sin \theta_2$$

and

$$\theta_3 = \sin^{-1}\left(\frac{a - b \sin \theta_2}{c}\right)$$

The velocities and accelerations are obtained by differentiating the equation for θ_3 as follows:

$$\omega_3 = \frac{d\theta_3}{dt}; \ \alpha_3 = \frac{d^2\theta_3}{dt^2}$$

or

$$\omega_3 = \frac{- b \cos \theta_2 \omega_2}{c\sqrt{1 - \left(\frac{a - b \sin \theta_2}{c}\right)^2}} = \frac{- b \ \omega_2 \cos \theta_2}{c \cos \theta_3}$$

while

$$\alpha_3 = -\frac{b}{c}\left[\frac{(\omega_2 \cos \theta_2)' \cos \theta_3 - \omega_2 \cos \theta_2 (\cos \theta_3)'}{\cos^2 \theta_3}\right] =$$

$$-\frac{b}{c}\left[\frac{(\alpha_2 \cos \theta_2 - \omega_2^2 \sin \theta_2) \cos \theta_3 + \omega_3 \omega_2 \cos \theta_2 \cos \theta_3}{\cos^2 \theta_3}\right] =$$

$$\alpha_3 = \frac{b \ \omega_2^2 \sin \theta_2 - b \ \alpha_2 \cos \theta_2 + c \ \omega_3^2 \sin \theta_3}{c \cos \theta_3}$$

The calculations for $\dot{x}$, and $\ddot{x}$ are:

$$x = b \cos \theta_2 + \sqrt{c^2 - (- b \sin \theta_2 + a)^2}$$

Then

$$\dot{x} = \frac{dx}{dt} = - b\omega_2 \sin \theta_2 + \frac{2 (a - b \sin \theta_2) b \ \omega_2 \cos \theta_2}{2 \sqrt{c^2 - (a - b \sin \theta_2)^2}}$$

Using

$$\sin \theta_3 = \frac{a - b \sin \theta_2}{\sin \theta_3}$$

yields

$$\dot{x} = \frac{b\omega_2 \sin \theta_2 \cos \theta_3 + b\omega_2 \sin \theta_3 \cos \theta_2}{\cos \theta_3}$$

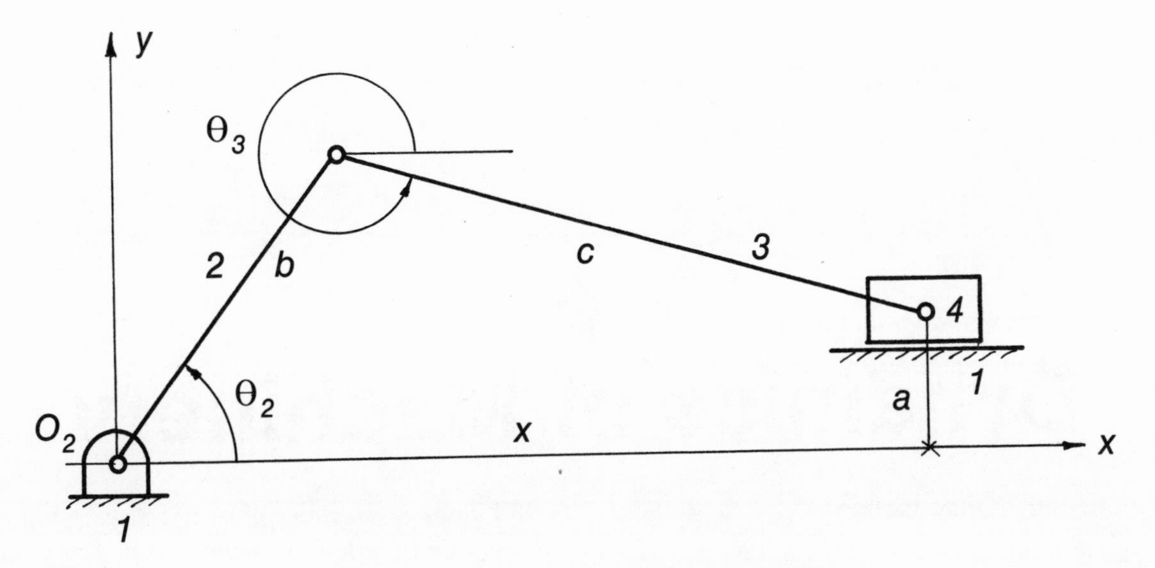

Figure 1-8. *Offset-slider crank mechanism.*

or

$$\dot{x} = \frac{b\omega_2 \sin(\theta_3 - \theta_2)}{\cos\theta_3}$$

Also

$$\dot{x} = b\omega_2 \tan\theta_3 \cos\theta_2 - b\omega_2 \sin\theta_2$$

Therefore

$$\ddot{x} = \frac{d\dot{x}}{dt} = b\alpha_2 \frac{\sin\theta_3}{\cos\theta_3}\cos\theta_2 + b\omega_2\omega_3 \frac{\cos\theta_2}{\cos^2\theta_3} -$$

$$b\omega_2^2 \frac{\sin\theta_2 \sin\theta_3}{\cos\theta_3} - b\alpha_2 \sin\theta_2 - b\omega_2^2 \frac{\cos\theta_2 \cos\theta_3}{\cos\theta_3}$$

$$\ddot{x} = \frac{b\alpha_2 \sin(\theta_3 - \theta_2) - b\omega_2^2 \cos(\theta_3 - \theta_2) - c\omega_3^2}{\cos\theta_3}$$

Most basic kinematics books cover this case and contain development of similar problems. These and other equations (for inversions of the four-bar linkage) were obtained in a different manner (by use of complex numbers) as shown in reference 1. In fact, the use of complex numbers may be preferable because of simplicity.

2 Dynamics of Machinery

Introduction

Machines are designed to accomplish a useful task, and they must work efficiently. The work done is a result of the action of forces or torques. Machine elements must be designed so they will be capable of carrying such loading. There are different kinds of loading; for example, static and dynamic forces. All input loads are balanced by resisting forces or torques.

Forces in mechanisms are determined in much the same way as in structures by use of graphical statics or analytical methods. The terminology used in statics (free-body diagram, equilibrium, two- and three-force members) also applies to the force analysis of mechanisms. An example of a statically determined structure is a kinematic chain with zero degrees of freedom. However, there may be a difference in loading. In structures, external forces are usually applied at the joints, while in mechanisms, the external forces are applied at points other than joints causing bending moments.

Links in mechanisms, or parts of machines, which are acted upon by external loads (forces and torques) will transmit the loads in different ways. Forces and motion are transmitted by a direct contact at a point, by a flexible link (belt), or otherwise. Two distinct situations should be observed when working dynamic problems dealing with link-to-link pair connections. These situations are distinguished by the presence or absence of friction. Friction will be insignificant in well-lubricated contacts because the force transmitted will be normal to the surface of contact. For example, in a lower-revolute-pair, a pin joint, the resultant force passes through the center of the joint. But in higher pairs, cams and gears, a line contact takes place and force action is normal to the contact surfaces. In spur gears, the force acts along the line of action at a pressure angle ϕ. In other types of gears, helical, bevel, or worm gears, the line of force transmission is more complicated due to the geometry of engagement.

All links in mechanisms will behave according to Newton's second law of motion when acted upon by external loads.

Using the principle of d'Alembert, this moving or accelerated system can be thought of as being brought to the state which is often called dynamic equilibrium. In fact, the d'Alembert principle is merely a rearrangement of Newton's second law ($F = ma$, Euler's equation $T = I\alpha$) resulting in an apparently simpler equation of motion ($F - ma = 0$; $T - I\alpha = 0$).

The advantage found in using d'Alembert's principle over the use of Newton's equation is that since the body is at "rest" you may take summation of moments (torques) about any axis. In Newton's equation, the action of forces and moments had to be considered about the center of gravity axis. Also, by using d'Alembert's equation or principle of adding a fictitious inertia force and couple to the acting external loads, you may reduce the number of equations required to solve the problem by eliminating unwanted unknowns. For example, a rolling disk on an incline requires two of Newton's equations of motion, $m\ddot{x} = mg \sin \alpha - F$ and $I\alpha = FR$, while only one d'Alembert equation is needed ($mgR\sin\alpha - I\alpha - m\ddot{x}R = 0$) if you take moments about the point of contact between the rolling body and incline. d'Alembert's principle offers a conceptual advantage in setting up the equations of motion; its use is a matter of preference.

Since a mechanism is a system of several rigid bodies, it is necessary to set up the equations of motion of each body and treat them as a coupled system of equations.

As in kinematic analysis (velocities, accelerations), graphical techniques are being used for solving vector equations for determining forces represented by directed line segments. Forces and torques are vectors. Solutions of equations containing forces or torques will be accomplished by vector algebra, and any polygon closure will simply mean the solution to an equation has been obtained.

Graphical methods have advantages of providing greater insight into the force-torque analysis and are often shorter and less complicated than the analytical methods. For a high degree of accuracy and an extension of the analysis for any position of the mechanism during the whole cycle of motion, an analytical method must be used.

By drawing what are called "Free-Body Diagrams," the forces of separate components can be analyzed in a more straightforward way. The system of rigid bodies is split into simple links, each represented by a free-body diagram. Using this technique, the next step is to solve equations that result from equilibrium conditions as known in statics.

Free-body diagrams give the opportunity of collecting and observing all the forces involved step-by-step. When drawing free-body diagrams, certain important rules must be observed:

1. Make sure that all the forces are accounted for and included
2. Place a force at every point of contact between two links
3. Direct the forces properly depending upon the type of contact involved (lower or higher pairs) or with or without friction

In a cam or slide contact, the forces should be normal to the surface if no friction is present. When friction is included, the frictional force should be tangent to the surface, and in the direction opposite to impending motion. When dealing with journal bearings, direct all forces to the center of rotation if no moment is present.

Notation

To avoid misunderstanding when referring to various forces, they will be denoted by two integer subscripts. The first integer indicates the number of the link providing the load, and the second integer indicates the link on which it is applied. Example: F_{12} would mean the force of link 1 on link 2, while F_{21} would be read as the force of link 2 on link 1.

$$F_{12} + F_{21} = 0 \text{ or } F_{12} = -F_{21}$$

In general, all links would be subject to the action of a number of forces and the following distinction is common:

1. Two-force members
2. Three-force members
3. More than three-force members

Examples of applications are shown in Figures 2-1 through 2-6. Numerous examples of application of two-force members exist. A connecting rod of a piston engine would be one example of application when the weight and inertia force are neglected, because the only locations where external forces are applied are at the crank and at the wrist pins. Moreover, these two forces must be equal in direction and opposite in sense. In tension, for example, they are directed toward each other. Now, if an external moment is applied to a two-force member, the two forces will have to be offset a distance "h" which would be calculated from the formula $h = M/F$. Other examples of application of two-force members are structural trusses, belts, ropes, and rods subject to tension and compression.

As an example of application of a three-force member, consider the link shown in Figure 2-1 having three forces applied to it. For equilibrium, the lines of action of the three forces must be either concurrent or parallel. If they are not parallel, their vector force-triangle must be closed. Since for equilibrium the summation of moments about any point must be zero, the force $\vec{F}_{13}$ must be directed toward point "P".

Point "P" is located at the intersection of the lines of action of the two remaining forces, $\vec{F}_{43}$ and $\vec{F}_{23}$. Note that if an external moment is applied to the body, the line of action of the arbitrarily selected third force will be offset from the intersection of the two forces by a distance of $h = MF$. Figures 2-2 and 2-3 show other examples of three-force members.

By adding an additional external force or an inertia force to a three-force member, you will obtain members of the third group—more than three-force members. On the other hand, you can do the opposite, reduce a more than a three-force member to a three-force member by adding two of the four or more forces together and the analysis proceeds as before. Examples of application for such a case are shown in Figures 2-5 and 2-6.

Sometimes a situation occurs where the lines of action of forces at the constraints (supports, joints) are not easily determined because none of the links are joined by two-force members. A method of superposition is quite common in mechanism force analysis. The basic idea behind this method is that the results of different forces can be superimposed or added separately. Fortunately, the relationship is linear—the results are independent of each other and can be added. That simplifies the procedure for solving such a problem. For example, consider a static-dynamic force analysis of a certain link of weight, W, acted upon by an external load P. You may disregard the inertia effect, solve for the pin reactions due to the external load P, and then repeat the calculations disregarding the weight, P, but considering the inertia force only. Finally, add the results obtained

(Text continued on page 10)

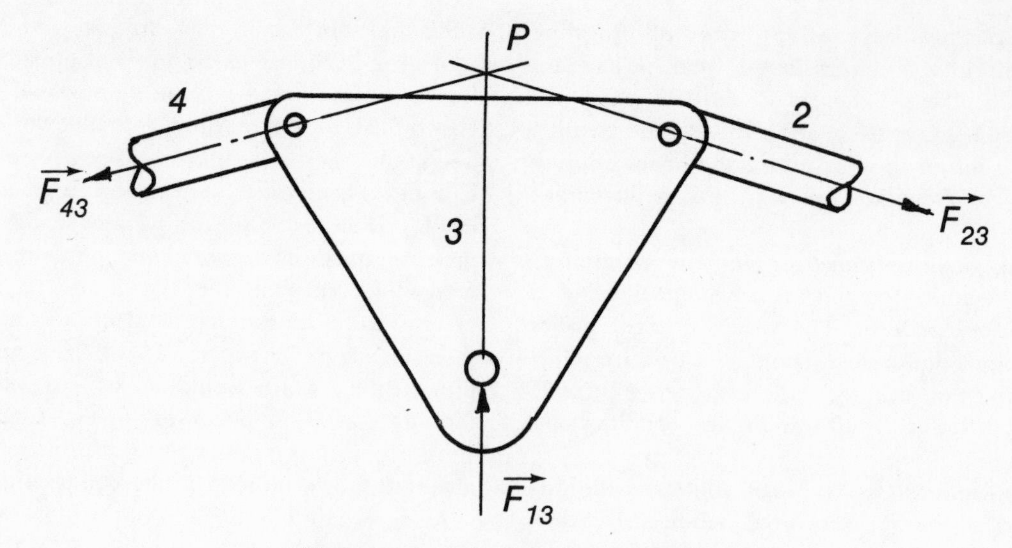

Figure 2-1. *A three-force member.*

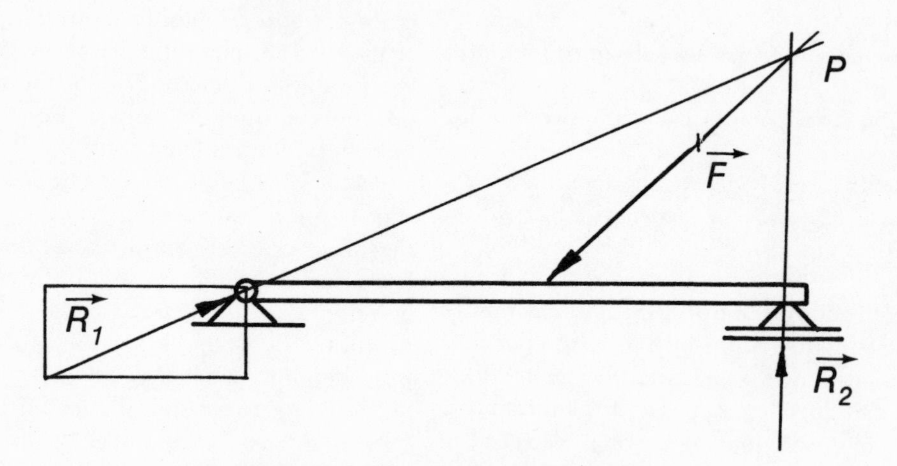

Figure 2-2. *A three-force member.*

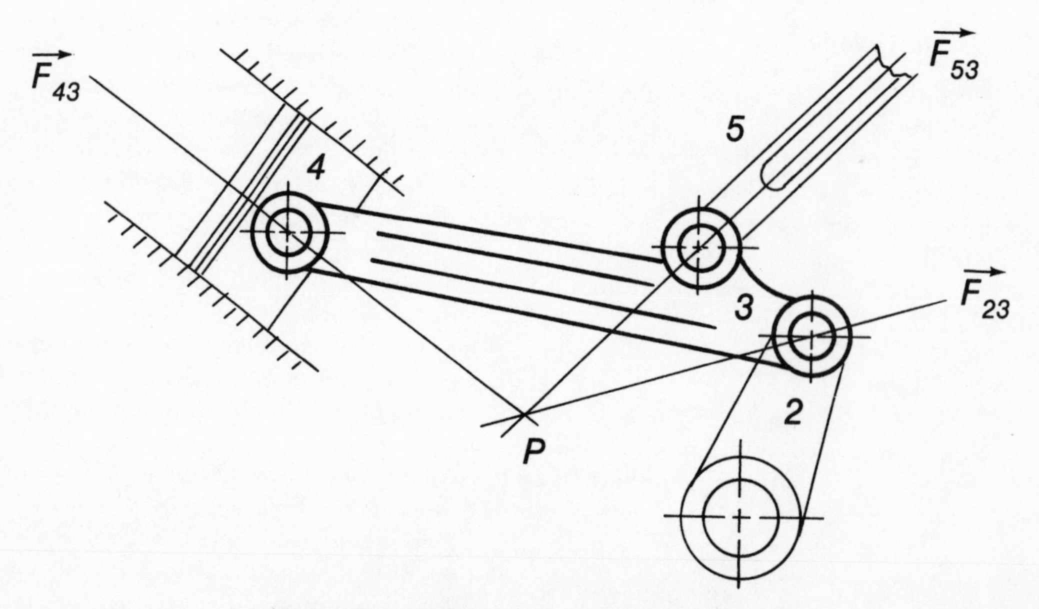

Figure 2-3. *A three-force connecting rod.*

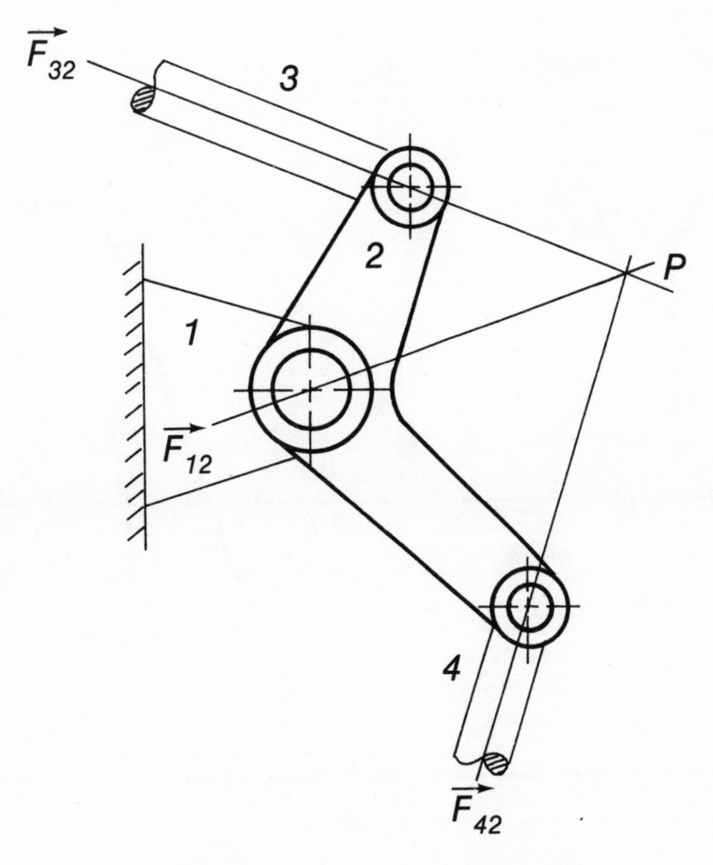

Figure 2-4. *A three-force bell crank.*

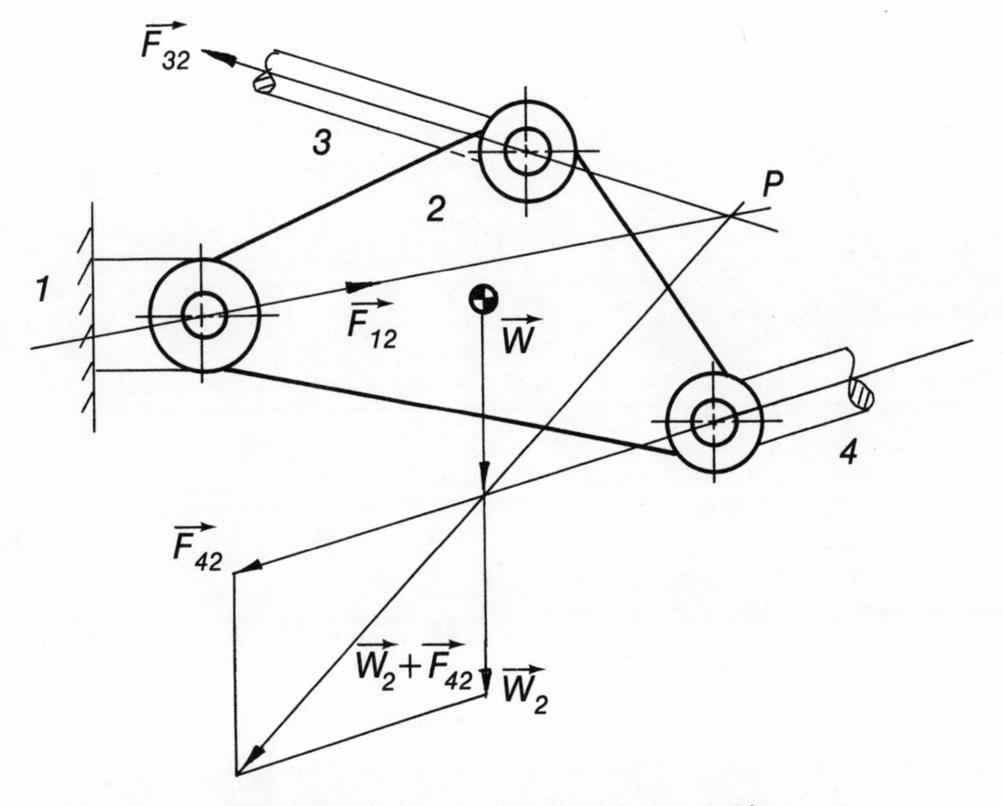

Figure 2-5. *A more than three-force member.*

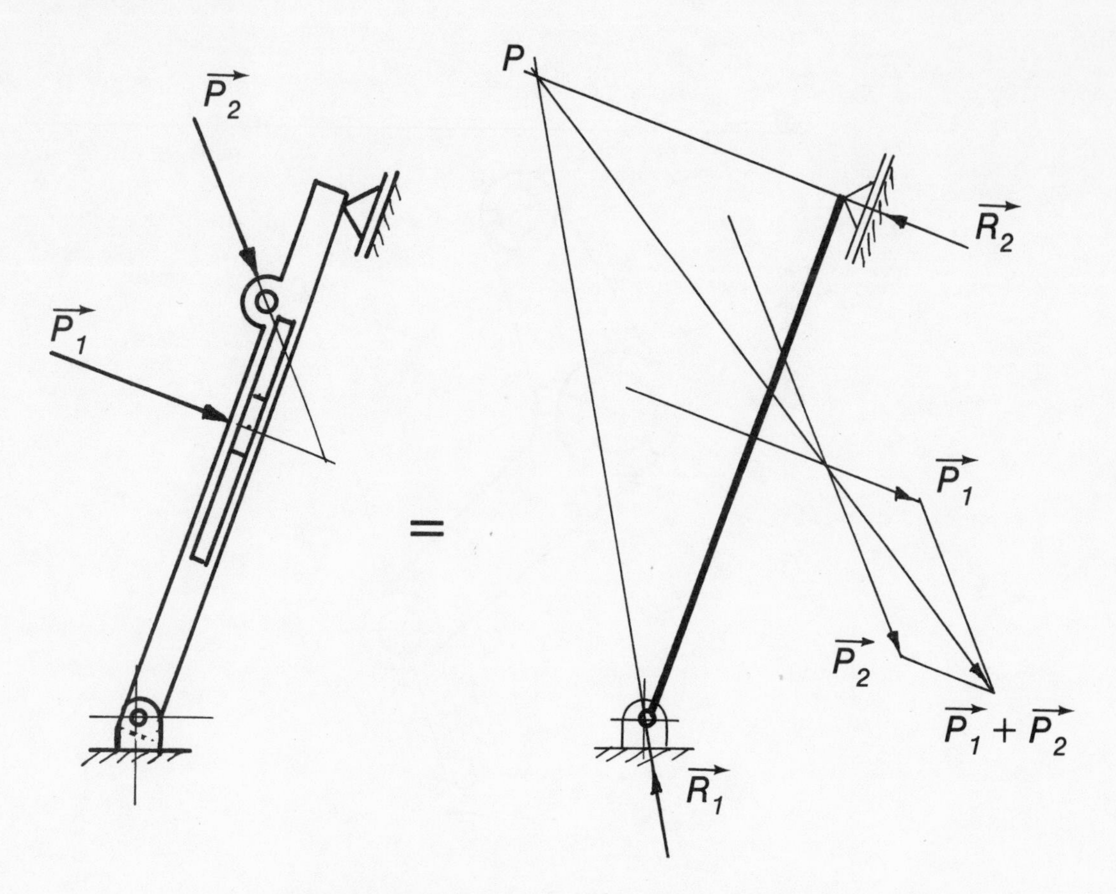

Figure 2-6. *A more than three-force member.*

in each step. This is the heart of the superposition method (Figure 2-13).

There are techniques other than those described in this text, graphical and analytical, and they are discussed and presented in extensive literature on this subject (e.g. pole-force method).

Direct Use of Newton's Laws

Really, how is the dynamics problem approached in general? In the following paragraphs, you will see that you are quite handicapped unless computers or other means of generating and solving equations of motion are used. The direct use of Newton's laws is questionable except for the simplest problems.

Take for example the five-bar, two-degrees of freedom mechanism shown in Figure 2-7. The equations of motion could be formulated by applying Newton's laws directly to each link of the linkage shown. The figure shows a set of the four free-body diagrams. The reaction forces at the joints, the torques, and the moments are also shown. Assuming that the torque T_2 is the given driving torque, there are then ten unknowns (F_{12x}; F_{12y}; F_{23x}; F_{23y}; F_{34}; T_A; F_{45x}; F_{45y}; F_{15} and T_B) plus four unknown variables defining the position (r_B, θ_2, r_{AB} and θ_4) resulting in a total of 14 unknown quantities.

For a planar system, you can obtain 3 equations of motion for each link of the free-body diagrams for a total of 12 equations. The other equations must be obtained from the geometry or the constraints of the mechanism. To show how complex this problem would become under these conditions, the equilibrium equations for some of the links and constraints equations are shown.

Link 2

$$\Sigma F_x = m\ddot{x}_2;\ F_{12x} + F_{23x} = m_2 \frac{d^2}{dt^2}\left(\frac{l_2}{2}\cos\theta_2\right)$$

$$\Sigma F_y = m\ddot{y}_2;\ F_{12y} + F_{23y} = m_2 \frac{d^2}{dt^2}\left(\frac{d_2}{2}\sin\theta_2\right)$$

$$\Sigma M_G = \bar{I}\alpha_2;\ T_2 + (F_{12x} - F_{23x})\frac{l_2}{2}\sin\theta_2 +$$

$$(F_{23y} - F_{12y})\frac{l_2}{2}\cos\theta_2 = \bar{I}\alpha_2$$

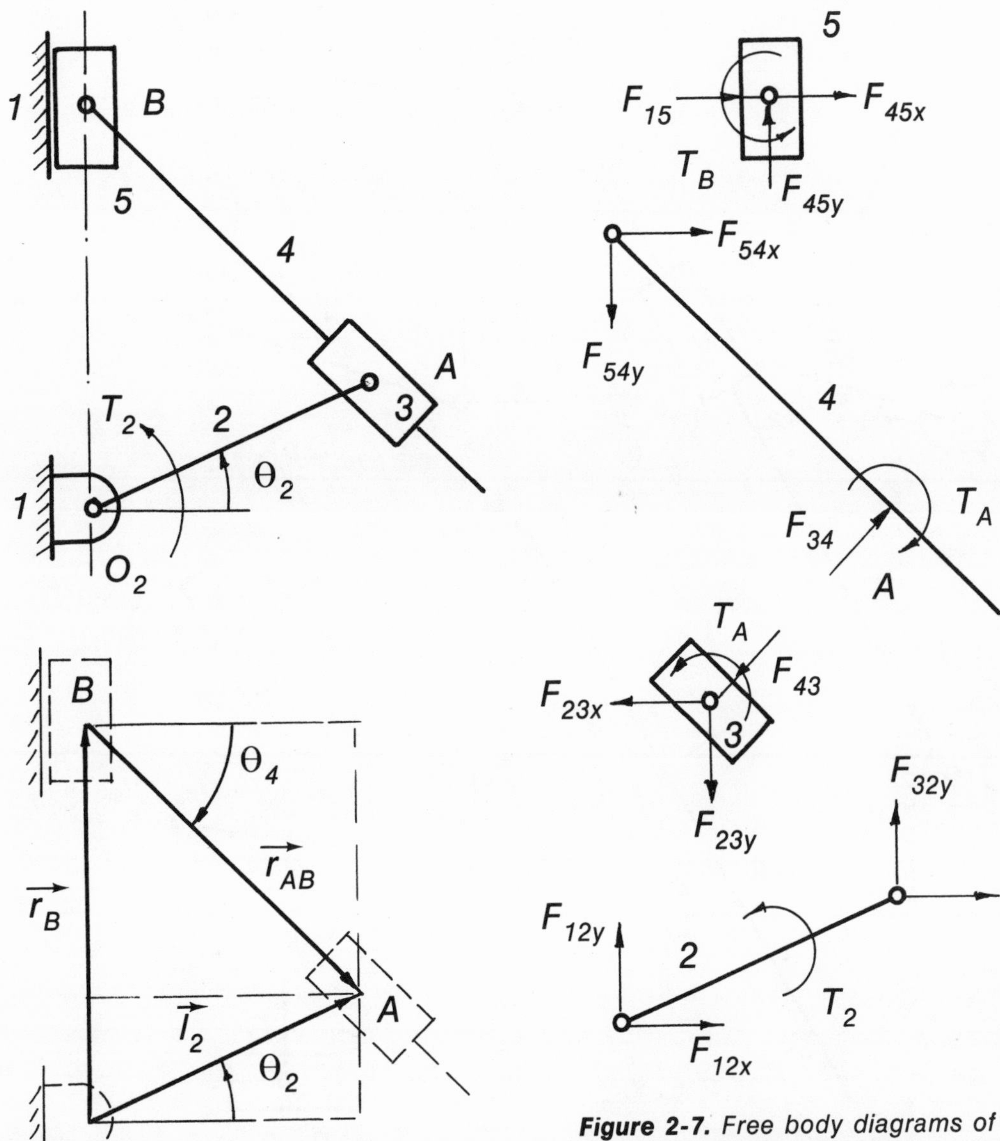

Figure 2-7. *Free body diagrams of a five-bar, two degrees of freedom linkage.*

Link 3

$$F_{23x} + F_{43} \cos (90 - \theta_4) = m_3 A_{3x}$$

$$F_{23y} + F_{43} \sin (90 - \theta_4) = m_3 A_{3y}$$

$$T_A = \bar{I}_3 \alpha_3$$

where $\bar{I}_i$ – moments of inertia about the mass centers at the midpoints of the links, $\alpha = \dfrac{d^2\theta}{dt^2}$ angular accelerations. Also, $F_{23x} = -F_{32x}$, etc.

The additional six equations for links 4 and 5 are obtained in a similar manner. The equations of constraints are obtained from the closure of the loop OABO in the form:

$$r_{AB} \cos \theta_4 + l_2 \cos \theta_2 = 0$$

$$r_{AB} \sin \theta_4 - l_2 \sin \theta_2 - r_B = 0$$

Solving this system of 14 equations for every position of the crank is a task for a computer. So, the direct application of Newton's laws to force analysis of machines is not feasible. Some of the existing programs (MEDUSA) use this approach, but they use methods of reducing the equations and shortcuts. This problem would be even more complex for the six-bar quick return mechanism shown in Figure 2-8. Twenty equations are needed. Figure 2-8 is provided so that you can write out the equations as a review.

Is there then a method for avoiding the tedious procedure resulting from the direct application of Newton's

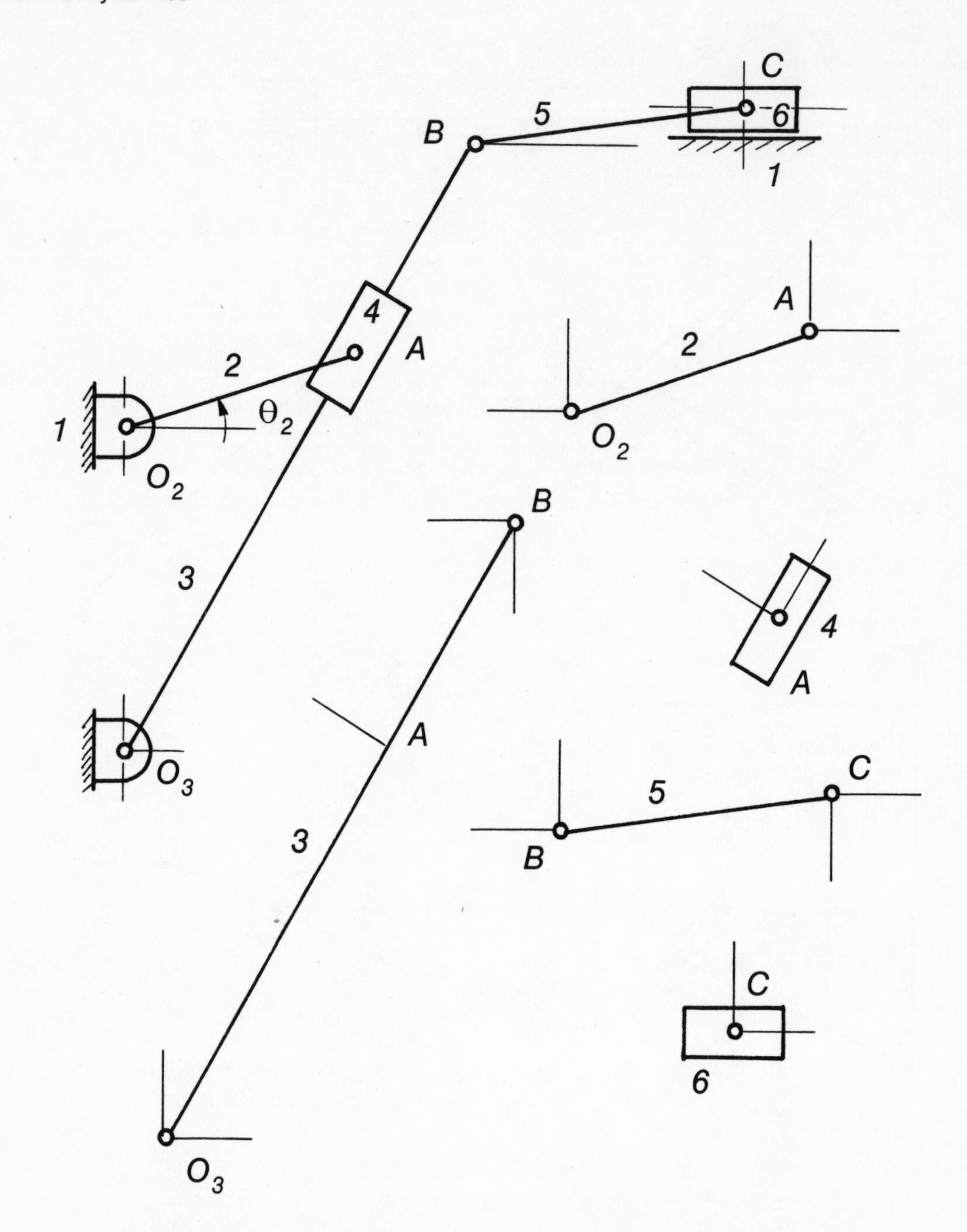

Figure 2-8. *Free body diagrams of a quick return mechanism.*

laws? The answer is yes, and some of the different methods of doing these types of problems are using advanced dynamics techniques.

d'Alembert's Principle

Dynamics of machinery problems are solved in two ways—as "kinetostatic" or as "time response" problems. When the motion of the mechanism is given (all link dimensions, angles, velocities and accelerations are known), the objective of the kinetostatic analysis is to determine reactions, forces, and torques acting at different joints and links of the mechanism. Additionally, the shaking forces and shaking moments may be determined.

The solution will be written in the form of a system of algebraic equations by using the equilibrium conditions for each link. The "time response" approach is a more complicated one, whenever the input force (torque) is given as a function of time, and the objective is to determine the motion characteristics (velocity, acceleration).

First, examine the shovel mechanism shown in Figure 2-9. Next, consider a two-degree of freedom planar five-link mechanism, which will be a more difficult one due to the more complex form of equations of dynamic equilibrium for each link.

Some assumptions made for the example problems are:

1. All dimensions are given
2. All links are considered to be rigid bodies
3. Joints are considered frictionless
4. Masses and moments of inertia of links are known
5. T_2 is the only external load acting on the driver link 2

Other forces and torques could be considered as well, if known. d'Alembert's principle is used when writing the equilibrium equations. (Figures 2-10 and 2-11.)

The dynamic equilibrium equations for each of the moving links of the shovel mechanism are given in the following sections.

Link 2

$$\Sigma F_x = 0;\ F_{12x} + F_{32x} = 0$$

$$\Sigma F_y = 0;\ F_{12y} + F_{32y} = 0$$

$$\Sigma T_{02} = 0;\ F_{32x} l_2 \sin\theta_2 - F_{32y} l_2 \cos\theta_2 - T_2 = 0$$

Link 3

$$\Sigma F_x = 0;\ -F_{32x} + F_{43x} - (m_3 A_{G3x}) = 0$$

$$\Sigma F_y = 0;\ -F_{32y} + F_{43y} + (m_3 A_{G3y}) = 0$$

$$\Sigma T_A = 0;\ F_{43x} l_3 \sin\theta_3 - F_{43y} l_3 \cos\theta_3 -$$

$$(m_3 A_{G3x}) \frac{l_3}{2} \sin\theta_3 - (m_3 A_{G3y}) \frac{l_3}{2}$$

$$\cos\theta_3 + (\bar{I}_3 \alpha_3) = 0$$

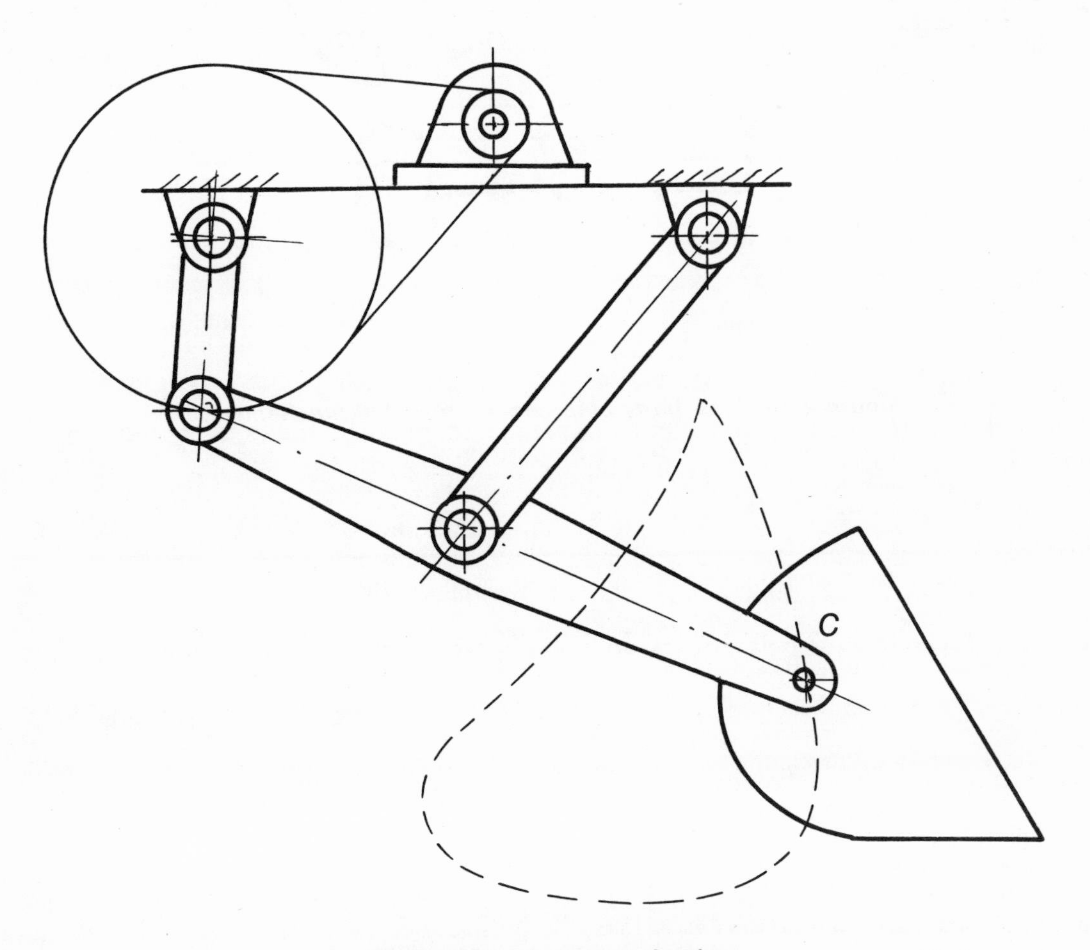

Figure 2-9. *Shovel mechanism.*

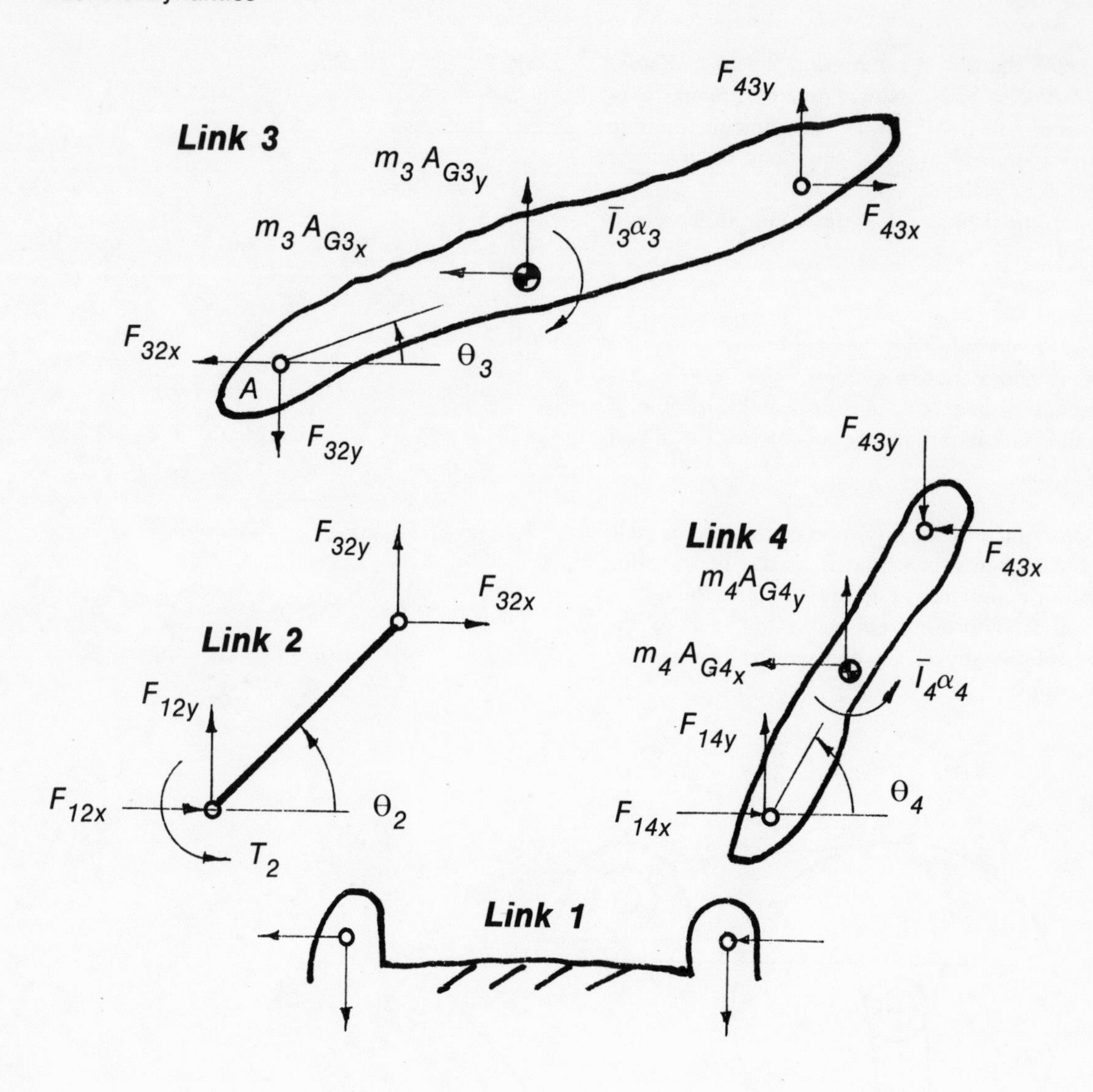

Figure 2-10. Free body diagrams of a shovel mechanism.

Link 4

$$\Sigma F_x = 0;\ F_{14x} - F_{43x} - (m_4A_{G4x}) = 0$$

$$\Sigma F_y = 0;\ F_{14y} - F_{43y} + (m_4A_{G4y}) = 0$$

$$\Sigma T_{04} = 0;\ -F_{43x}l_4\sin\theta_4 + F_{43y}l_4\cos\theta_4 - (m_4A_{G4x})\frac{l_4}{2}\sin\theta_4 - (m_4A_{G4y})\frac{l_4}{2}\cos\theta_4 - (\bar{I}_4\alpha_4) = 0$$

Substituting into the above equations $\theta_2 = 150°$; $\theta_3 = 29.44°$; $\theta_4 = 116.18°$

$l_2 = 3''$; $l_3 = 8''$; $l_4 = 6''$

$m_3A_{G3x} = 152.57$ lb

$m_3A_{G3y} = 70.82$ lb.; $m_4A_{G4x} = 66.52$ lb; $m_4A_{G4y} = 18.79$ lb; $\bar{I}_3\alpha_3 = 139.78$ in.-lb; $\bar{I}_4\,\alpha_4 = 82.68$ in.-lb

yields:

$$F_{12x} + F_{32x} = 0$$

$$F_{12y} + F_{32y} = 0$$

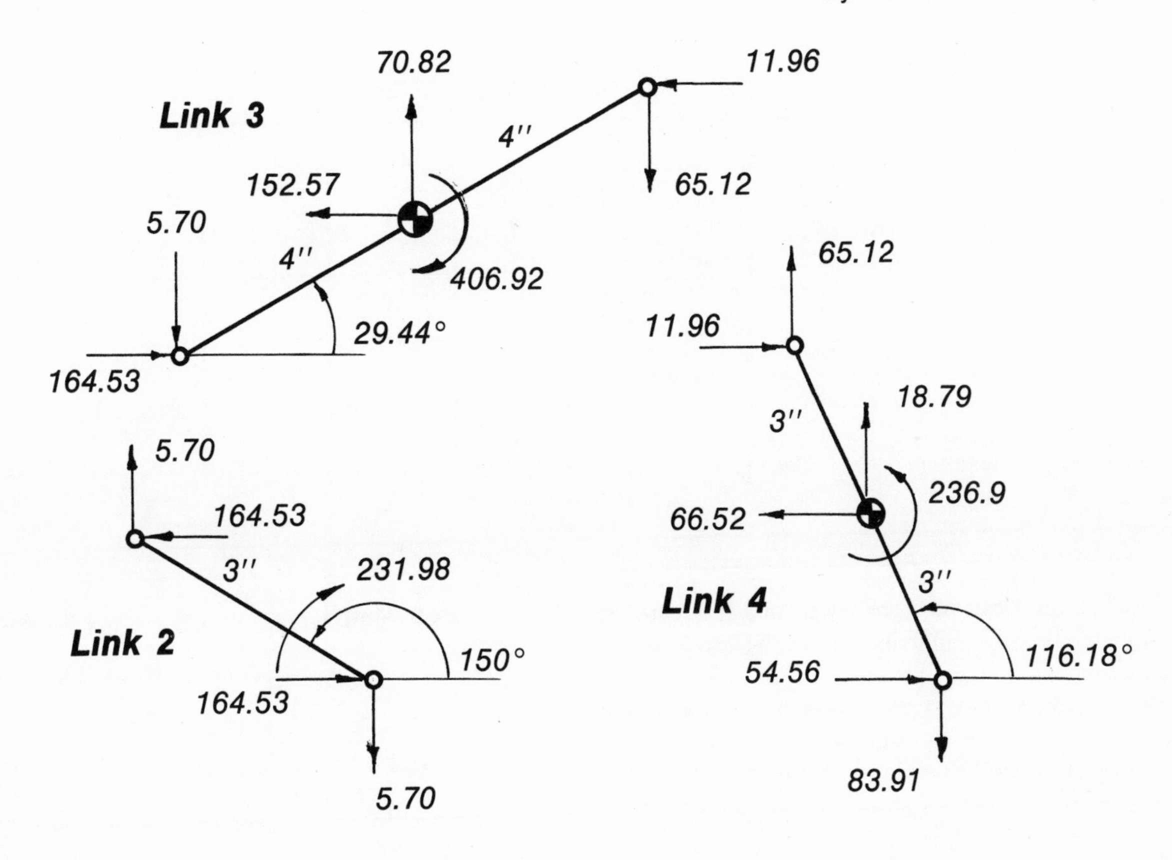

Figure 2-11. *Link equilibrium check.*

$$1.5\ F_{32x} + 2.6\ F_{32y} - T_2 = 0$$

$$-\ F_{32x} + F_{43x} - 152.57 = 0$$

$$-\ F_{32y} + F_{43y} + 70.82 = 0$$

$$3.93\ F_{43x} - 6.97\ F_{43y} - 406.92 = 0$$

$$F_{14x} - F_{43x} - 66.52 = 0$$

$$F_{14y} - F_{43y} + 18.79 = 0$$

$$-\ 5.38\ F_{43x} - 2.65\ F_{43y} - 236.90 = 0$$

The system of equations will be set up in the form of a matrix as shown in Table 2-1.

As a result, a system of nine linear algebraic equations with nine unknowns (T_2, F_{12x}, F_{12y}, F_{23x}, F_{23y}, F_{34x}, F_{34y}, F_{14x} and F_{14y}) has been written. The system of equations must be solved to obtain the external torque T_2 required to produce the given motion (velocities and accelerations are already known).

Solving a system of nine equations can be done by use of a pocket calculator without difficulty. In fact, the TI 59 calculator handles a nine-by-nine matrix in about 12 minutes.

Taking the shaking force as the vector sum of all the inertia forces yields:

$$F_{sx} = -\ m_2 A_{G2x} - m_3 A_{G3x} - m_4 A_{G4x}$$

$$F_{sy} = -\ m_2 A_{G2y} - m_3 A_{G3y} - m_4 A_{G4y}$$

or

$$F_s = \sqrt{F_{sx}^2 + F_{sy}^2}$$

The shaking moment, which is by definition the sum of all the inertia couples and moments due to the inertia forces about an arbitrary point O of the frame, will be:

$$M_{sO2} = T_2 + F_{41y}\ a$$

$$M_{sO4} = T_2 + F_{21y}\ a$$

Table 2-1
Force Matrix

F_{32x}	F_{32y}	F_{43x}	F_{43y}	T_2	F_{12x}	F_{12y}	F_{14x}	F_{14y}	
1.5	2.6	0	0	−1	0	0	0	0	0
−1	0	1	0	0	0	0	0	0	152.57
0	−1	0	1	0	0	0	0	0	−70.82
0	0	3.93	−6.97	0	0	0	0	0	406.92
0	0	−5.38	−2.65	0	0	0	0	0	236.90
1	0	0	0	0	1	0	0	0	0
0	1	0	0	0	0	1	0	0	0
0	0	−1	0	0	0	0	1	0	66.52
0	0	0	−1	0	0	0	0	1	−18.79

These equations can be solved repetitiously for any desired increase of the input angle θ_2 of the crank, and the results obtained can be presented graphically. This will allow investigation of the maximum torques, shaking force, shaking moments, and bearing reactions during the complete cycle of motion of the mechanism.

Note:

1. The directions of the inertia forces and torques as shown in Figure 2-10 have been determined for the crank position of 150° by kinematics on pages 29 and 30.
2. The TI-59 Master Library module has a previously mentioned capability of solving for the determinant or inverse of a 9 × 9 matrix, but it can not solve a set of a system of 9 equations with 9 unknowns, so two runs were required.
3. An elegant and powerful method for solving this problem is the so called "matrix method" where the geometry coefficients and dynamical terms are arranged in a form of a matrix and solved easily by a computer.

Procedure for use of a TI-59 calculator:

1. Repartition memory to 159.99 (Press 10, 2nd OP 17)
2. Select program (Press 2nd, PGM 02)
3. Enter order of matrix
4. Calculate determinant
5. Solve equations

These results check with both the graphical and analytical solution. The printouts are:

5 x 5 Matrix

5.
1.5
−1.
0.
0.
0.
2.6
0.
−1.
0.
0.
0.
1.
0.
3.93
−5.38
0.
0.
1.
−6.97
−2.65
−1.
0.
0.
0.
0.

Δ= 47.9131

0.
152.57
−70.82
406.92
236.9

−164.5261247 F_{32x}
5.696962668 F_{32y}
−11.95612473 F_{43x}
−65.12303733 F_{43y}
−231.9770842 T_2

For $m_4 = 0$

0.
152.57
−70.82
406.92
0.

−130.0638795 F_{32x}
25.12832904 F_{32y}
22.50612046 F_{43x}
−45.69167096 F_{43y}
−129.7621638 T_2

For $m_3 = 0$

0.
0.
0.
0.
236.9

−34.46224519 F_{32x}
−19.43136637 F_{32y}
−34.46224519 F_{43x}
−19.43136637 F_{43y}
−102.2149203 T_2

7 x 7 Matrix

Δ = 47.9131

0.
152.57
−70.82
406.92
236.9
0.
0.

−164.5261247 F_{32x}
5.696962668 F_{32y}
−11.95612473 F_{43x}
−65.12303733 F_{43y}
−231.9770842 T_2
−164.5261247 F_{12x}
5.696962668 F_{12y}

0.
152.57
−70.82
406.92
236.9
66.52
−18.79

164.5261247 F_{32x}
5.696962668 F_{32y}
11.95612473 F_{43x}
65.12303733 F_{43y}
231.9770842 T_2
54.56387527 F_{14x}
83.91303733 F_{14y}

Using optimization techniques, a four-bar linkage has been selected (dimensions were obtained from a synthesis of such a mechanism) from an infinity of possible combinations of link proportions.[2]

The different optimum solutions resulted in reduction of the shaking forces, shaking moments, and bearing forces up to 80 percent respectively. Therefore, the dynamic performance has been greatly improved.

Of course, such a procedure would be a very difficult task without the use of a computer. For the second example, the problem would be even more difficult if direct Newton's equations (and sum of moments) were written without determining the velocities and accelerations first, and then substituting them into the equilibrium equations as data. The resulting equations are quite complicated, as seen in the example.

A more convenient way of solving these equations is by use of rectangular components $\vec{i}$, $\vec{j}$, and $\vec{k}$. This is shown in detail in the "Shovel Mechanism" section of this chapter.

Graphical Dynamic Analysis

Whenever the effects of motion are included in force analysis, a dynamic force analysis takes place. *Static force analysis* means determination of forces or reactions at the joints and torques with the assumption that the inertia forces and couples are negligible. Motion of mechanisms depends on the geometric characteristic of the links, and it depends on the constraints of the linkage system. According to Newton's second law, in order to obtain accelerations of machine parts, forces and/or couples must be imposed on the system. Such loads result in motion and also produce stresses in machine elements. A review of the basics of dynamics will be carried for one case only showing how the couple and force can be replaced by a single equivalent force.

Imagine a simple link with two forces acting on it as shown in Figure 2-12A. The center of gravity of this link is found to be at point G. Adding the two forces together vectorially will determine the magnitude and direction of the resultant force $\vec{R}$. The effect of the resultant force will be a linear acceleration $\vec{A}_G$ of the link plus a clockwise angular acceleration $\vec{\alpha}$. Consider the same link with the addition of a reversed force called an inertia force equal to $m\vec{A}_G$, and also with a reversed couple, $I\vec{\alpha}$, called an inertia couple as shown in Figure 2-12B.

Notice that the inertia force $m\vec{A}_G$ passes through the center of gravity G and is equal, but of opposite sense (not direction) to $\vec{R}$. Being at a distance "d" from the center G, the force $\vec{R}$ produces a torque "$\vec{d} \times \vec{R}$" about G. With the inertia force and inertia couple, the link will be in a state which is often called *dynamic equilibrium*. This permits us to use statics techniques. Taking the summation of moments about G, we find $\vec{d} \times \vec{R} = \vec{I} \times \vec{\alpha}$. The next step is to replace the inertia force and inertia couple by a single force. Such a substitution is not required for an analytical solution, but to use a force polygon (graphical solution), a conversion of the three-dimensional force-torque system to a planar system is necessary. The representation of the inertia force and inertia couple by a single force will simplify the solution because the two inertia effects will be taken care of by a single vector instead of two vectors. This procedure is shown in Figure 2-12C.

It is possible to replace the couple by any other couple composed of two equal and opposite forces of any magnitude and direction as long as the distance "h" between the two forces is such that "$\vec{h} \times \vec{F}$" equals "$\vec{I} \times \vec{\alpha}$". Notice in Figure 2-12C that the shifting of the equivalent couple is such that one of the forces of the equivalent couple is along the same line of action as the inertia force, but of opposite sense. Since the equal and opposite forces cancel, a single force remains. The distance "h" is calculated from the formula:

$$h = \frac{I_G \alpha}{m A_G}$$

Note: The line of action of the single force is shifted in a direction from the center of gravity G so that the moment of the single force "$\vec{h} \times \vec{F}$" about the center G will be in the same direction as the inertia couple and opposite to the angular acceleration itself. While the inertia vector is only an imaginery one, it does have some physical meaning as an indication of the resistance a mass offers when its velocity is changed with respect to time. While most of this material focuses on coplanar systems, force systems and motions in general are three-dimensional, and secondary inertia couples may also be present.

The objectives of dynamic analysis are:

1. Determination of forces and torques involved for a given motion
2. Determination of motion as a result of acting loads
3. Reaction forces, shaking forces, vibration, etc.

A step-by-step procedure of dynamic force analysis would most likely be as follows:

1. Make a schematic drawing of the mechanism with basic dimensions, joints, etc.
2. Locate the centers of gravity of all links which are moving (find $\bar{x}$, $\bar{y}$)

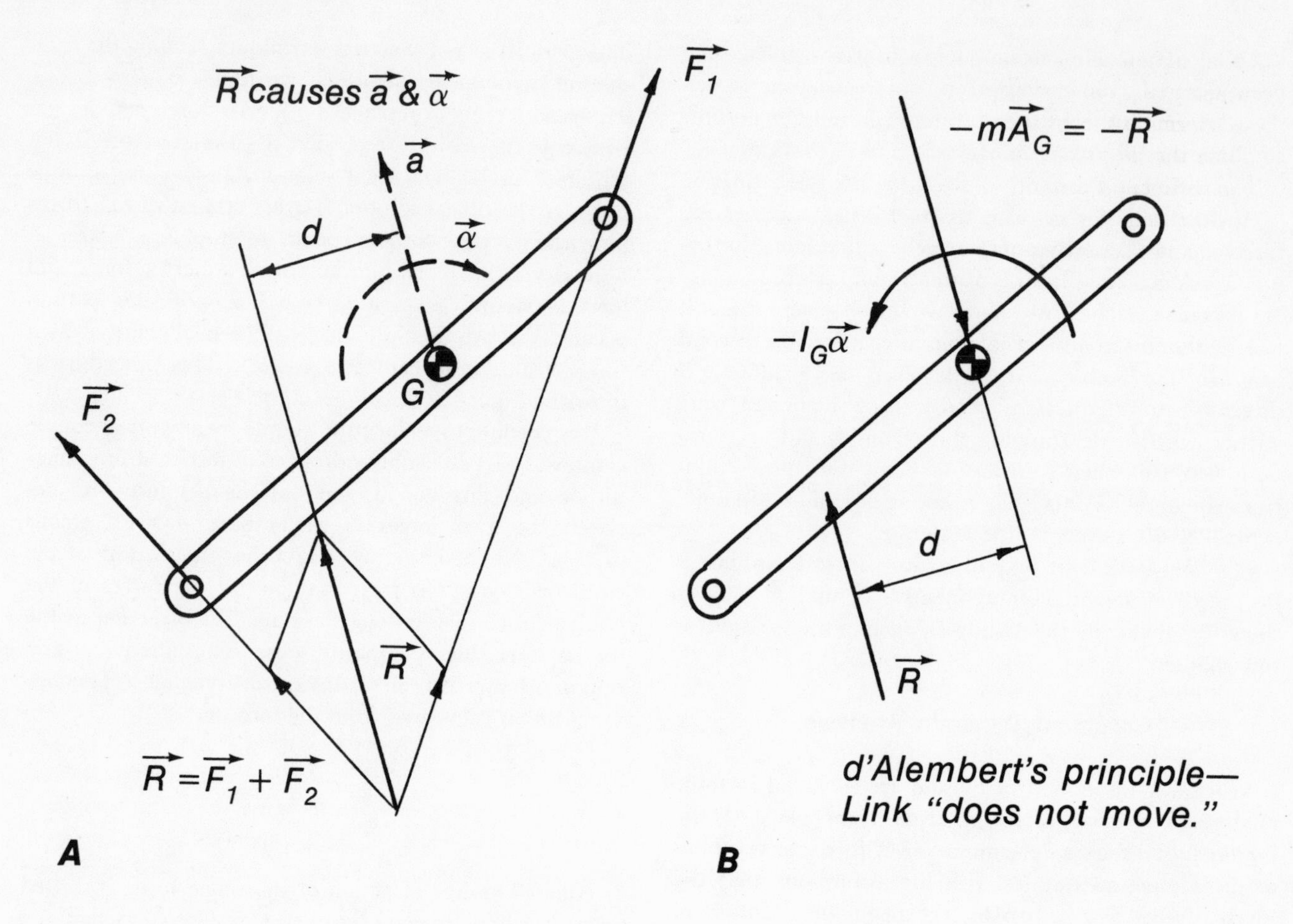

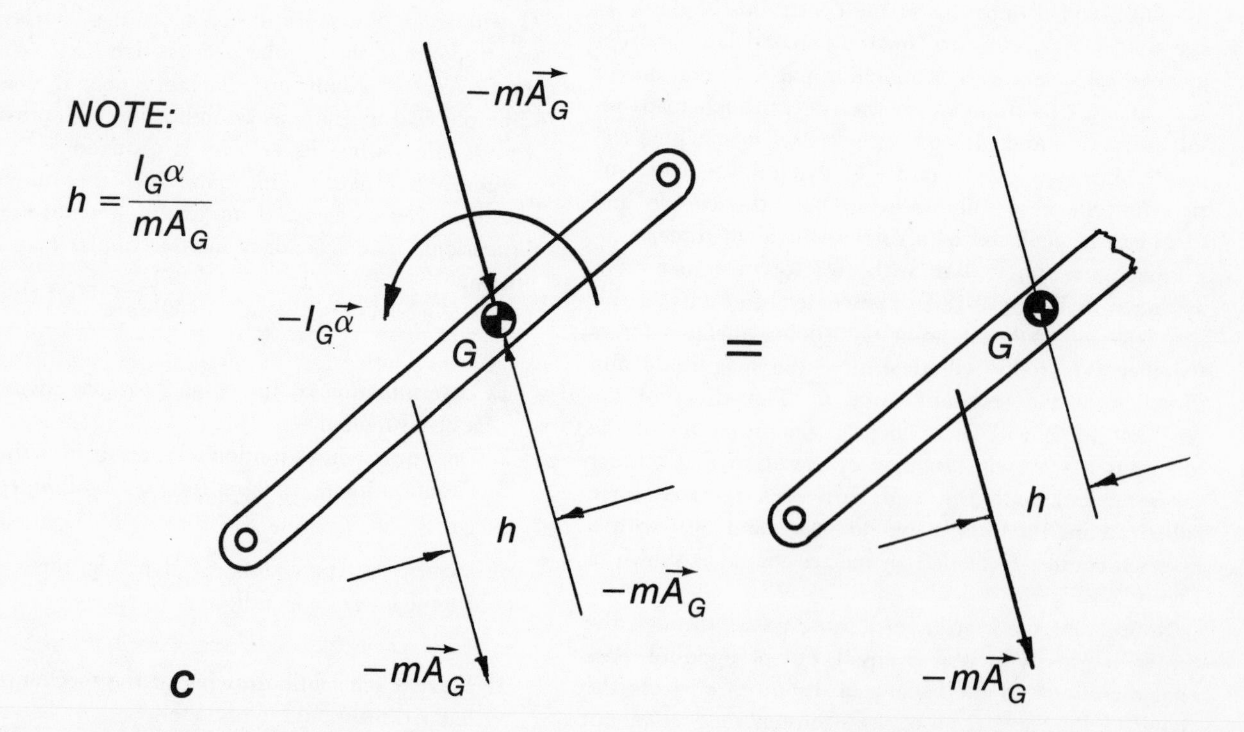

Figure 2-12. Free body diagrams of a link acted upon by two external forces.

3. Calculate the mass of each movable link (watch units)
4. Determine the moments of inertia needed to calculate the inertia effects using analytical formulas or experimental data
5. After drawing velocity and acceleration polygons (or analytically by using a program) determine the accelerations of the centers of gravity of the moving parts
6. Determine the inertia forces and couples $m\vec{A}$ and $\overrightarrow{\bar{I}\alpha}$
7. Locate the distance $h = \dfrac{\bar{I}\alpha}{mA_G}$ of the single force representing the inertia force and couple as described before (Make sure to apply it in such a direction that the inertia force aids the reversed inertia couple.)
8. Draw free-body diagrams for each member (link) separately and treat them as in statics, writing summation of forces and moments as equilibrium conditions and determining step-by-step forces needed
9. In analytical procedures skip part 7

As an example, consider the four-bar linkage shown in Figure 2-13 acted upon by two weights, $\vec{W}_3$, $\vec{W}_4$, and a force $\vec{P}$. The superposition method could be used in solving for the force $\vec{F}_{23}$ as shown in steps 1 through 6 in the figure.

Examples of *static force analysis* problems are shown in Figures 2-14, 2-15, 2-16, and 2-17.

Friction

The existence of friction complicates the solution of any dynamic problem, and in some cases neglecting friction would invalidate the solutions.

Friction forces are those which always oppose motion. They can be desirable or undesirable, depending on the machine. Generally, these resisting forces can be best defined by the following formula:

$$F = -k\,f(N, v, x)$$

where k is either μ (coefficient of friction) or c (coefficient of damping), N is the normal force component, v is the velocity, and x is the displacement.

Basically, there are two major frictional forces, static and sliding (i.e., Coulomb friction). Frictional forces at a surface, and couples at a hinge, could be calculated when values are known for μ, the normal force components, the hinge loads, and the dimensions. Bearing catalogues list the coefficients of friction μ while other engineering handbooks give values of μ for sliding. Therefore, the problem becomes one of obtaining the normal force components and the hinge torques. They can be realistically evaluated by different methods, as previously shown.

Unfortunately, only some very simple problems of force analysis including Coulomb friction could be solved by the use of the so-called friction circle, e.g., the slider crank.

It is probably best to completely ignore friction unless a more detailed, in-depth study of friction is to be conducted with the help of a computer. Extensive literature is available and recommended whenever a complete force analysis is required.

Since this is beyond the scope of this text, friction has been neglected in the preceeding force analysis examples (Figures 2-14, 2-15, 2-16 and 2-17).

Determination of Mass Moments of Inertia *I*

The mass moment of inertia I of a link about its center of gravity may be determined analytically (by use of basic formulas $I = \int r^2 dm$ and $I = \bar{I} + md^2$) or experimentally. It is not difficult to determine moments of inertia for homogenous bodies with simple geometrical shapes. But, for bodies of nonhomogenus materials or of complicated shapes, calculations may be impossible or inaccurate. Therefore, experimental procedures for finding moments of inertia are used. There are different methods for setting up experiments for determining the moment of inertia. One method measures the frequency of vibration, for example.

If the part which is to be used was available, or a model could be made, analytical calculations could be double-checked by experimental procedure described in literature on dynamics and vibration.

In all of the problems presented, assume that the weight and mass moment of inertia are given or could be determined, e.g., $W = 3$ lb and $\bar{I}_G = 0.003$ lb-sec^2-ft. When the linear and angular accelerations of the mass centers are determined, the inertia force will be known. The shaking force example of the Flying Shear Mechanism shows how to find the inertia moments step-by-step.

Shovel Mechanism

As an example, conduct a dynamic force analysis of a shovel mechanism lifting a load as sketched in Figure 2-9. More specifically, find the external torque which must be applied to the input link and the forces acting at each pin joint. Also, discuss ways of improving the torque characteristics if needed. Use data as given and $k_s = 1{:}20$ (Figure 2-18).

(Text continued on page 25)

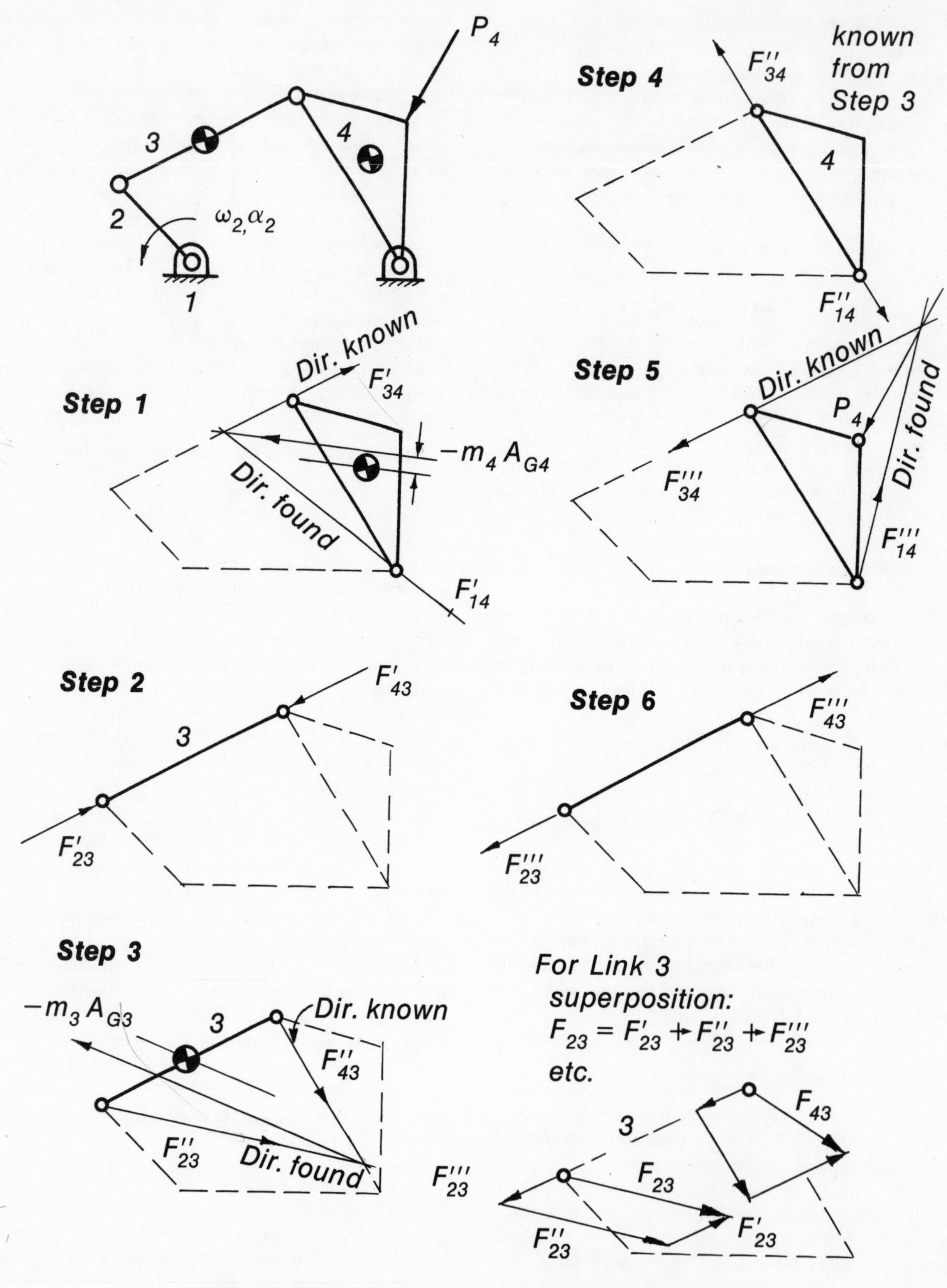

Figure 2-13. *Force analysis by the superposition method.*

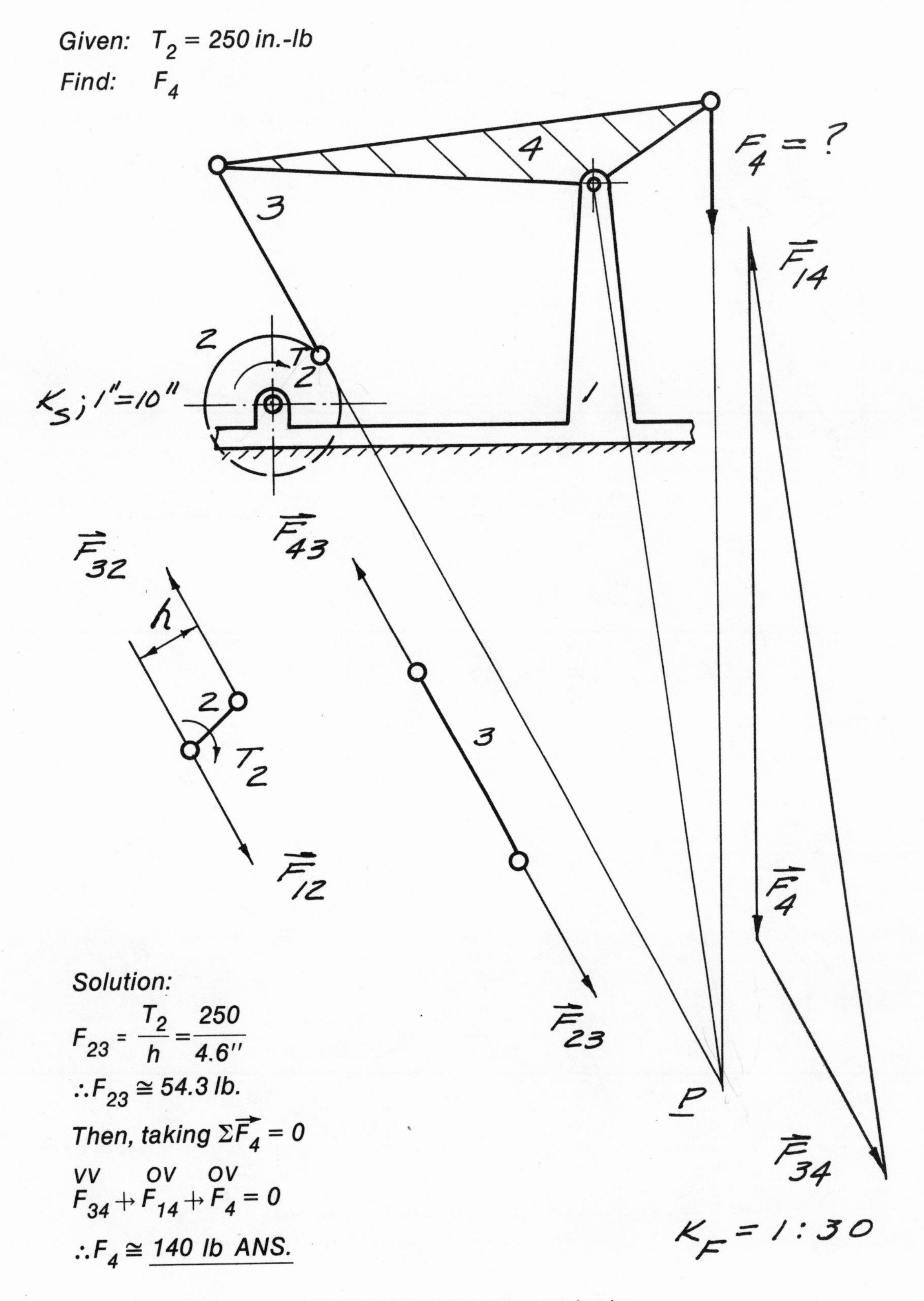

***Figure 2-14.** A four-bar mechanism.*

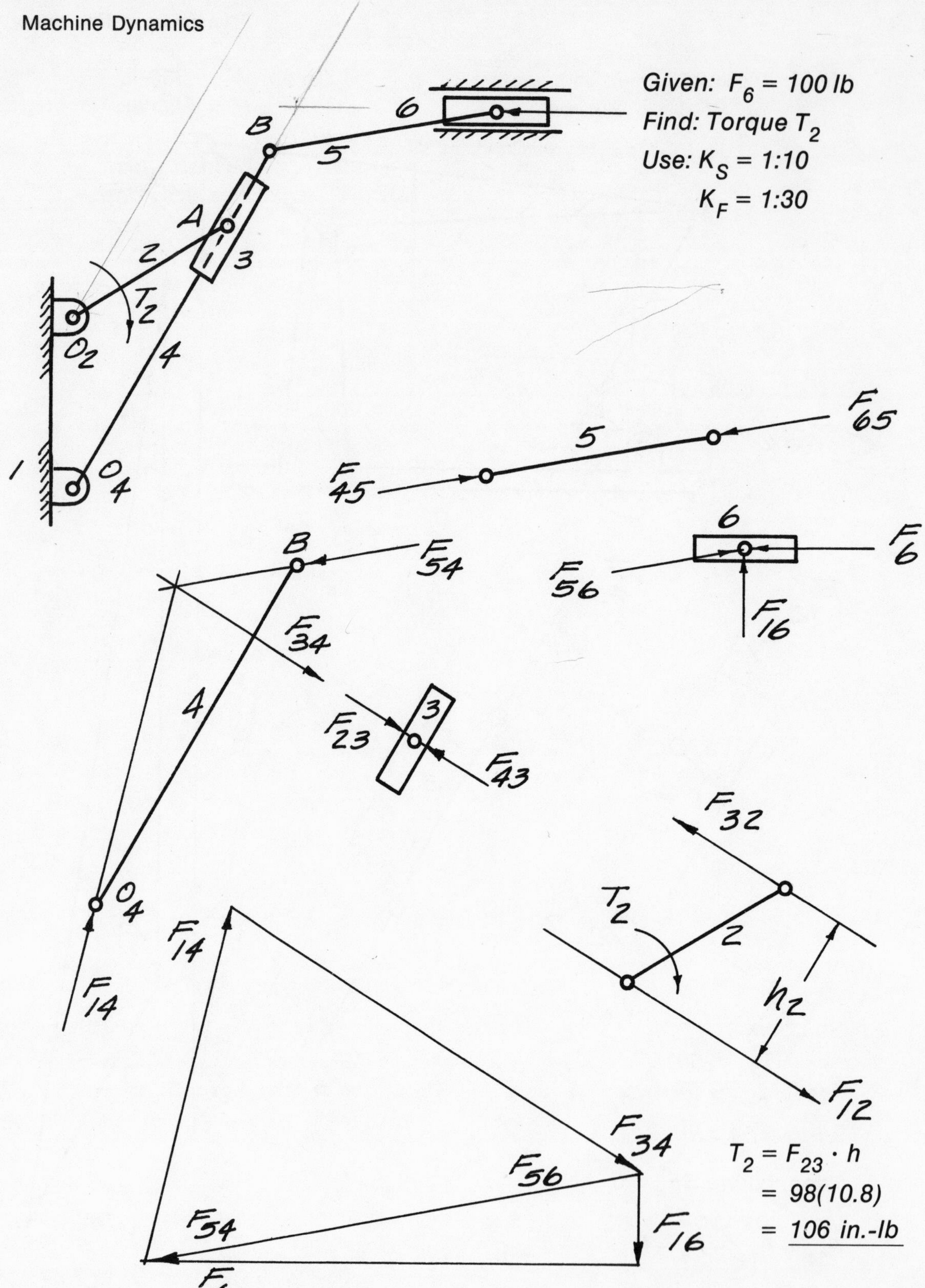

Figure 2-15. *A six-bar linkage.*

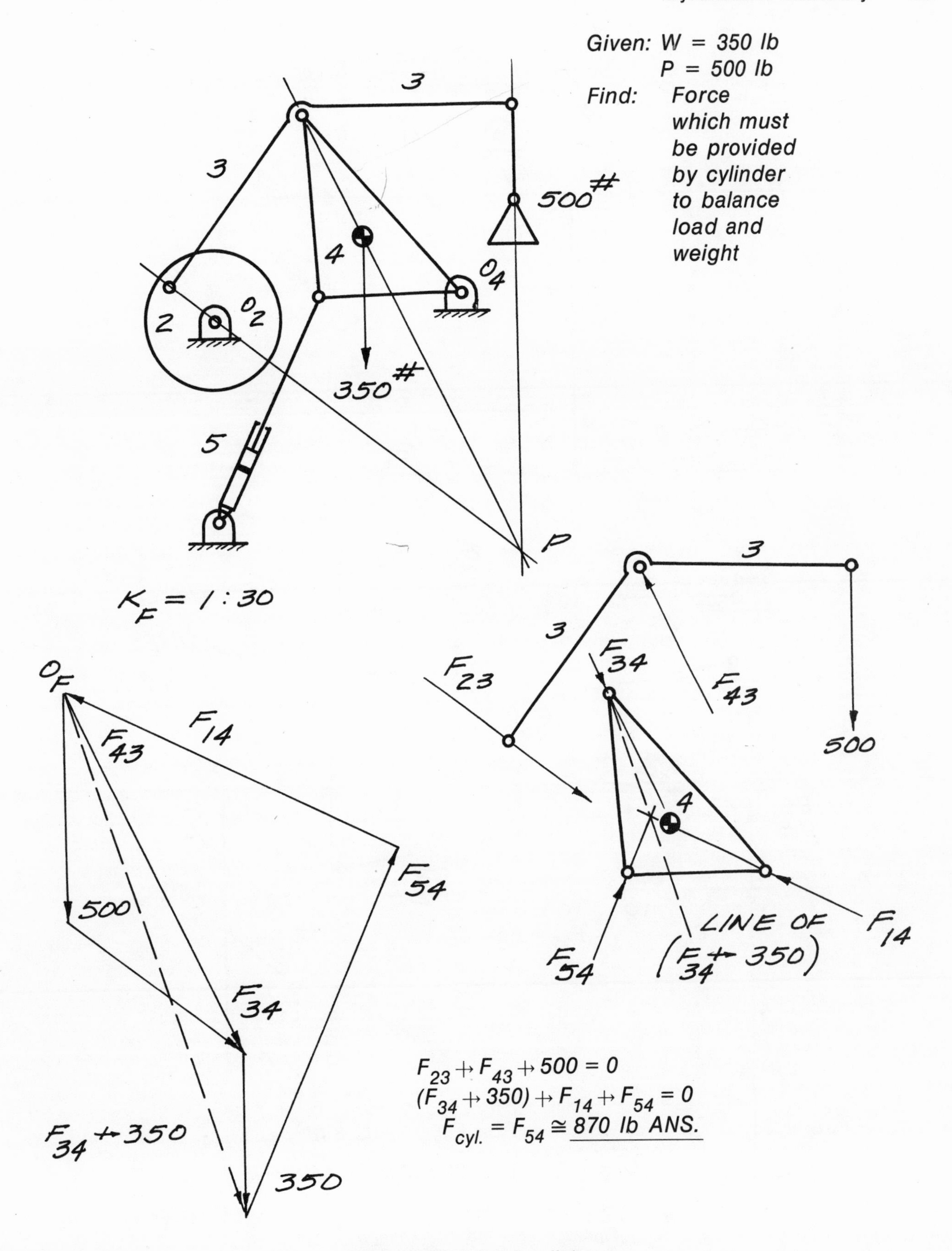

Figure 2-16. *A five-bar linkage.*

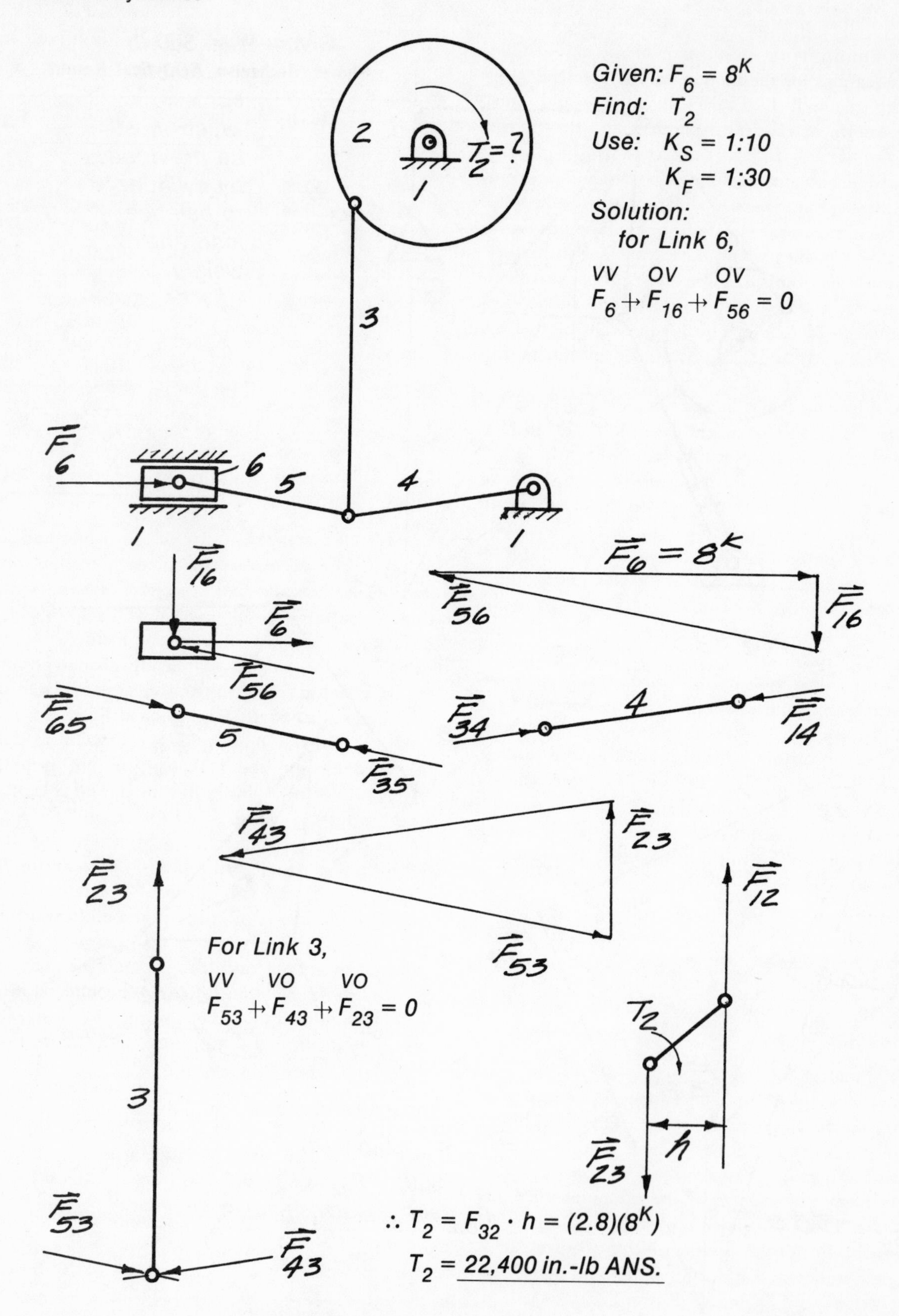

Figure 2-17. *A six-bar linkage.*

When conducting the dynamic motion analysis, general equations for the motion will be obtained (Virtual work method, Lagrange's equations, and other methods known in classical dynamics) and they should include the torque, T_2, external forces, F_3, F_4, spring forces, and linear and angular velocities and accelerations. A classical approach to this problem would result in a system of nonlinear equations quite difficult to solve analytically. Yet, they would have to be solved and the results would be graphed. From these graphs, a recommendation as to the overall feasibility and needed improvements would follow. This is somewhat beyond the scope of this text, as it is limited to determining and discussing the case of torque only.

These comments by no means complete the discussion of needed work for a complete dynamic study. In fact, other aspects such as vibration analysis (to avoid resonances), surging of springs, and impact loads or shaking forces, as done in the second example, should be taken into consideration. The task is a difficult one, but with the use of a computer program like the IMP Program, the problem could be solved much faster. Once again, it should be emphasized that without the help of the TI-59 calculator program included in Appendix 1, it would be very difficult to even obtain the torque versus position curve. Someone could probably question this approach and come up with a different one to find the torque. The problem is not a simple one. For example, not even the exact values for the external forces acting at point C are known. Therefore, another approach would be to use an electric motor and measure the current flow which would give the values for the input torque needed to run this machine. The objective here is to show how such a problem should be attacked and what needs to be done.

Solution Procedure

Three different methods were used in solving this problem:

1. A graphical solution for two values of the crank angle $\theta = 150°$ and $270°$ (Figures 2-18 and 2-19)
2. An analytical solution for one position, $\theta_2 = 150°$ using rectangular components as often done in basic dynamics courses
3. An analytical solution by use of complex numbers and a TI-59 calculator program (included in Appendix 1) for 12 positions, or for every 30° of crank rotation, with the assumption that link 4 is weightless (Table 2-2, Figure 2-20)

In all three solutions, $m_2 = 0$, and $m_3 \neq 0$.

There are advantages and disadvantages for each method used. Only a thorough review of each method

Table 2-2
Shovel Mechanism Analytical Results

θ_2	M_3AG_3	T_2	F_{23}	F_{34}
0°	404.45	−1125.17	613.04	334.77
10	450.00	− 848.43	734.09	480.18
30	330.94	111.82	508.09	306.23
60	179.62	224.97	197.39	21.98
90	166.73	51.34	107.61	− 59.27
120	168.64	− 67.16	105.86	− 68.71
150	168.20	− 130.08	132.80	− 50.80
180	170.59	− 126.72	159.01	− 26.69
210	174.18	− 68.17	172.50	− 12.12
240	177.73	15.52	180.12	− 12.81
270	179.49	143.65	197.87	− 38.39
300	158.38	417.98	239.99	−131.24
330	102.61	505.92	187.04	−266.76
350	281.00	− 706.14	313.86	48.84
360	404.45	−1125.17	613.04	334.77

will provide information as to which is best and should be used. The principles used in each method are the same. They are the basic tools for dynamics of machinery for this type of problem. Therefore, you must understand them. To master the methods, other examples can be worked by merely changing some of the data and checking the answer already given in the tabulated values of the third method listed.

Notice that the answers for the linkage, as done in the third method and tabulated, do not include the effects of inertia of link 4 due to the number of steps required for the TI-59 program used in this solution. By rewriting the program and by eliminating from it the numerous pauses, you can easily include the inertia of link 4 and solve the problem. The procedure for including the inertia of link 4 is straightforward, which gives the equations needed for the program.

Double checks of calculations should be made whenever there is a doubt as to the results obtained in either method. Although this problem is quite elementary, it is very easy to make a sign mistake and not realize that the solution is meaningless.

Analytical Solution

Given:

$\vec{\omega}_2 = +200\vec{k}$, rad/sec, ccw; $W_3 = 0.7$ lb; $W_4 = 0.8$ lb

$\alpha_2 = 0$; $\bar{I}_3 = 0.02$; $\bar{I}_4 = 0.01$ in.-lb-sec^2

$a = -7\vec{i} = 7\underline{/0°}$ in.

$b = -2.60\vec{i} + 1.50\vec{j} = 3\underline{/150°}$ in.

$c = 6.98\vec{i} + 3.94\vec{j} = 8\ \underline{/29.44°}$ in.

$d = -2.65\vec{i} + 5.39\vec{j} = 6\ \underline{/116.2°}$ in.

Comments About Assignments

The real art of any design is to present a complete solution, but is there such a thing as a complete solution? The dynamic analysis is merely a part of the design process, yet a very important part. Other aspects of the design process are: materials to be used and their properties, cost, technology of putting it all together, space, and maintenance. Overall, the combination of links acting together must perform a specific job. The shovel must move along a certain path obtained by synthesizing the coupler point curve *C*. (The Hrones and Nelson atlas would be the best source of curves and link proportions.)

You must choose the best linkage system you can find. A four-bar linkage is a good choice for such a machine, since it is the simplest, most useful, well defined and best known mechanism. As sketched, this mechanism is a crank-and-rocker one-degree of freedom linkage. Additional links may be needed for guiding the shovel or rotating it around point C. This could most likely be achieved by a force provided by a cylinder to balance the load and weight. To simplify, the analysis will be for the main skeleton of the shovel mechanism only.

How does a designer manage such a task of dynamic force analysis? In order to design a machine, you must begin with a prediction of its performance before the machine can be actually built. The prediction will be based on a mathematical description or model, and by applying physical laws of behavior of previously tested parts to the model. Part of the modeling is the assumption of dimensions of links, types of joints, number of links, and type of linkage system. The model will be a schematic representation, and it will be a compromise of simplicity versus accuracy. The results will be valid only if the model closely approximates the real physical machine. No schematic approximation will be precise, because estimates and assumptions must be made. Experiments would be useful in determining the validity of the assumptions made.

Any dynamic analysis will most likely require more than merely a force analysis. In fact, your task could be separated into several distinct parts, such as type synthesis and motion study, finding of velocity and acceleration, and a dynamic analysis including balancing and vibration considerations.

Therefore, in general, any problem similar to this one should be worked in the following manner:

1. Make schematic drawing (rough dimensions) of the mechanism including all links, joints, etc., and a motion study first
2. Using Newton's law or d'Alembert's principle, establish variables and equations for each link

More specifically, in this problem, you should choose the type of four-bar linkage (open or crossed), check for best transmission angle, determine velocities (angular and linear), and determine accelerations for specific points of interest (centers of gravity, etc.) for as many different positions of the input crank as needed. A computer program is almost a necessity in order to obtain a complete set of values for the input crank rotation of 360°. Next, a torque versus position graph and a pin reaction graph need to be plotted so that maximum values can be obtained. Double checks of calculations must be made to avoid errors. For example, the computer program could be easily checked by use of graphical methods for two or three different angle positions of the crank. If the solutions at any stage are not satisfactory, the mechanism must be redesigned by adjusting dimensions or even choosing a different linkage (a five-link mechanism or more). Finally, a prototype should be made and tested including balancing and other characteristics of the machine.

For zero speed, or some constant slow rotation of the input crank, it would be sufficient to design a linkage without considering the inertia effects just to make sure that it would perform the motion required. Static forces could be determined, and then the size of different links determined by using Strength of Materials formulas. It is unlikely that there will be a constant load acting on a linkage. In this particular problem, the load will be a variable one, and the machine will experience such a load when the shovel engages or releases the weight it will carry. This will have an effect on the input link because the load will only act for part of the rotation of the crank. With the help of a pocket calculator, you will be able to calculate through the whole cycle and plot the torque graph. This will require the determination of the inertia forces after the accelerations have been found. It is possible to balance the inertia effects for some conditions of the linkage system in the absence of external loads, but it will be more difficult to balance a variable external load by the use of springs (or other means) although this is possible.

Nevertheless, an in-depth study of the dynamic response is needed following the preliminary analysis of linkage geometry and kinematics.

Looking at the torque versus position graph for the linkage, observe the peak values. Measures should be taken to improve this by adding springs and reducing the peak values at different points. Similarly, a static-pin force analysis should be conducted to avoid bearing failures. As far as the dead weight of the links is concerned, it can be neglected depending upon its magnitude as compared with the other externally acting forces. The spring will also reduce the pin reactions if properly mounted.

(Text continued on page 29)

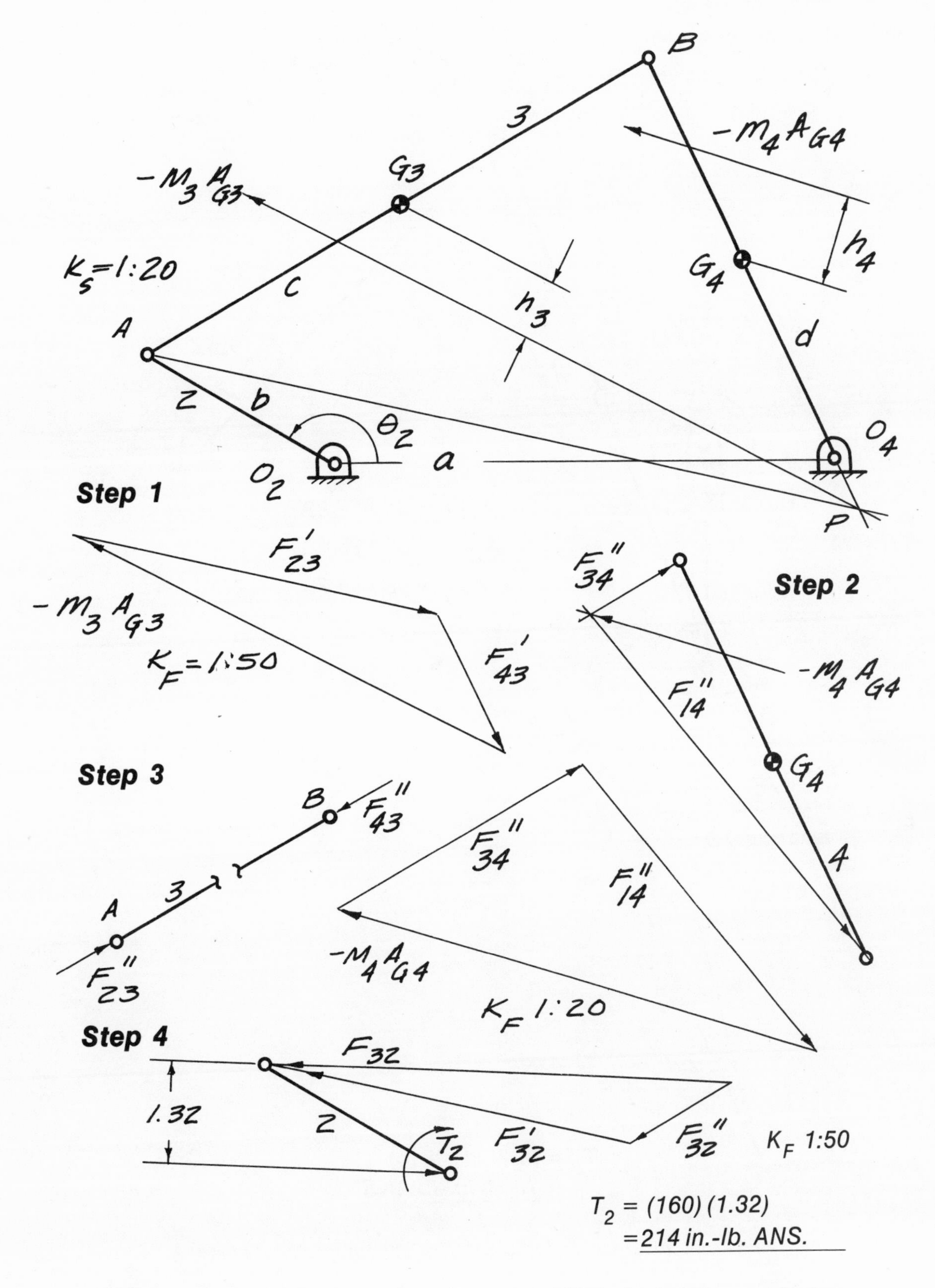

Figure 2-18. *Dynamic-force analysis for one position of a four-bar mechanism.*

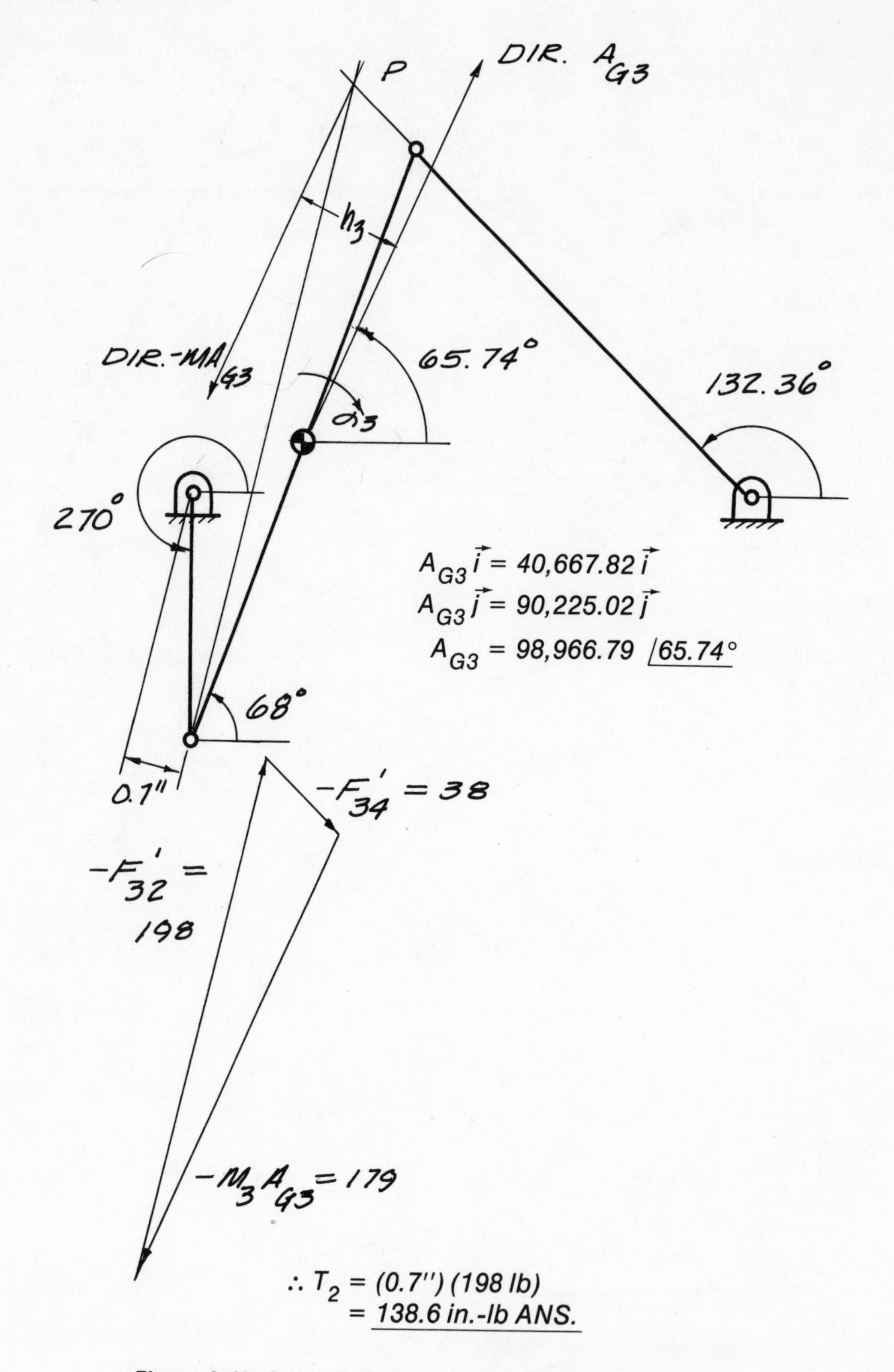

$\therefore T_2 = (0.7'')(198\ lb)$
$= \underline{138.6\ in.\text{-}lb\ ANS.}$

Figure 2-19. *Dynamic-force analysis without inertia of link 4.*

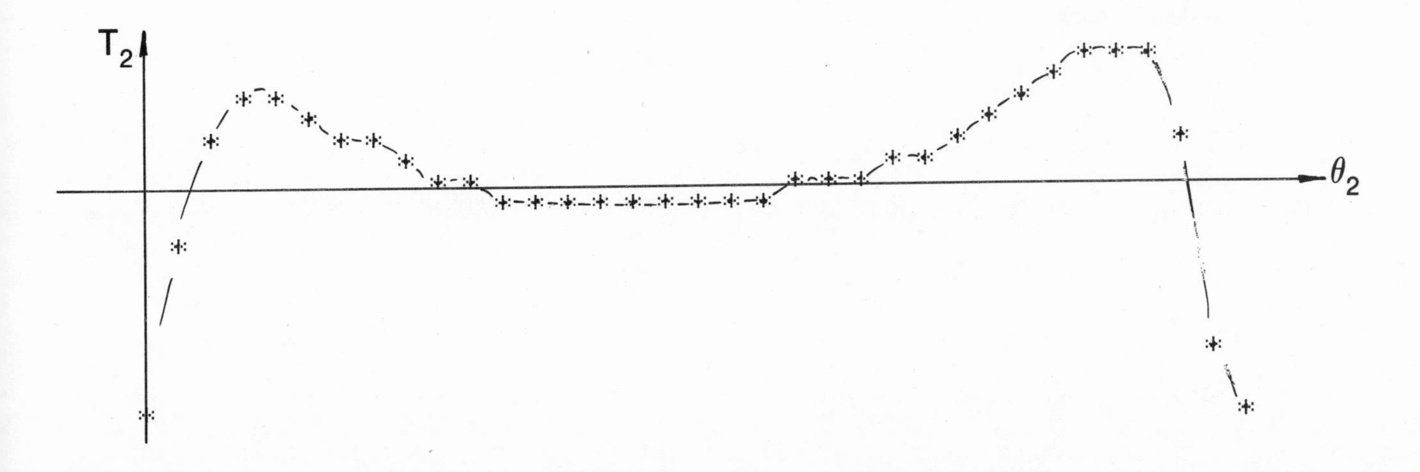

Figure 2-20. *Torque versus angular position diagram.*

Velocity and Acceleration Analysis

$\vec{V}_B = \vec{V}_A + \vec{V}_{B/A}$

$\vec{\omega}_4 \times \vec{d} = \vec{\omega}_2 \times \vec{b} + \vec{\omega}_3 \times \vec{c}$

$\omega_4\vec{k} \times (-2.65\vec{i} + 5.39\vec{j}) = 200\vec{k} \times (-2.61\vec{i} + 1.5\vec{j}) + \omega_3\vec{k} \times (6.98\vec{i} + 3.94\vec{j})$

i − comp. $\quad -5.39\ \omega_4 = -300 - 3.94\ \omega_3$

j − comp. $\quad -2.65\ \omega_4 = -520 + 6.98\ \omega_3$

$\therefore\ \omega_3 = +41.81$ rad/sec, ccw

$\omega_4 = +86.21$ rad/sec, ccw

$\vec{V}_B + \vec{\omega}_4 \times \vec{d} = -464.72\vec{i} - 228.48\vec{j} = 517.85\ \underline{/153.8°}$

$\vec{V}_{B/A} = \vec{\omega}_3 \times \vec{c} = -164.73\vec{i} + 291.83\vec{j} = 335.0\ \underline{/119.4°}$

$\vec{V}_A = -300\vec{i} - 520\vec{j} = 600\ \underline{/240°}$; $\vec{A}^t_A = 0$

$\vec{A}^n_B + \vec{A}^t_B = \vec{A}^n_A + \vec{A}^t_A + \vec{A}^n_{B/A} + \vec{A}^t_{B/A}$

$\vec{A}_A{}^n = 200\vec{k} \times (-300\vec{i} - 520\vec{j}) = 104{,}000\vec{i} - 60{,}000\vec{j} = 120{,}067\ \underline{/-30°}$

$\vec{A}^n_B = 86.21\vec{k} \times (-464.72\vec{i} - 228.48\vec{j}) = 19{,}697\vec{i} - 40{,}064\vec{j} = 44{,}644\ \underline{/-63.8°}$ in./sec^2

$\vec{A}^n_{B/A} = 41.81\vec{k} \times (-164.73\vec{i} + 291.83\vec{j}) = -12{,}201\vec{i} - 6{,}887.4\vec{j} = 14{,}010\ \underline{/209.44°}$ in./sec^2

$\vec{A}^t_{B/A} = \vec{\alpha}_3 \times \vec{c} = \alpha_3\vec{k} \times (6.98\vec{i} + 3.94\vec{j}) = -3.94\ \alpha_3\vec{i} + 6.98\ \alpha_3\vec{j}$

$\vec{A}^t_B = \vec{\alpha}_4 \times \vec{d} = \alpha_4\vec{k} \times (-2.65\vec{i} + 5.39\vec{j}) = -5.39\ \alpha_4\vec{i} - 2.65\ \alpha_4\vec{j}$

$19{,}697\vec{i} - 40{,}064\vec{j} - 2.65\ \alpha_4\vec{j} - 5.39\ \alpha_4\vec{i} = 104{,}000\vec{i} - 60{,}000\vec{j} - 12{,}201\vec{i} - 6{,}887\vec{j} - 3.94\ \alpha_3\vec{i} + 6.98\ \alpha_3\vec{j}$

i − comp. $\quad -72{,}102 = -3.94\ \alpha_3 + 5.39\ \alpha_4$

j − comp. $\quad 26{,}823 = 6.98\ \alpha_3 + 2.65\ \alpha_4$

$\alpha_3 = +6{,}989$ rad/sec^2, ccw

$\alpha_4 = -8{,}268.4$ rad/sec^2, cw

$\vec{A}_{G3} = \vec{A}_A + \vec{A}_{G3/A} = \vec{A}_A + (1/2)\vec{A}_{B/A}$

$\vec{A}_{G3} = 104{,}000\vec{i} - 60{,}000\vec{j} + 1/2[-12{,}201\vec{i} -$

$6{,}887\vec{j} + 6{,}989\ (6.98)\ \vec{j} - 3.94\ (6{,}989)\ \vec{i}] =$

$84{,}131\vec{i} - 39{,}052\vec{j} = 92{,}753\ \underline{/-24.89°}\ \text{in./sec}^2$

$\vec{A}_{G4} = \vec{\alpha}_4 \times \vec{r}_{G4} + \vec{\omega}_4 \times (\vec{\omega}_4 \times \vec{r}_{G4})$

$= -8{,}268\vec{k} \times 1/2\ \vec{r}_4 + 86.21\vec{k} \times$

$(86.21\vec{k} \times 1/2\ \vec{r}_{G4})$

$= 32{,}129\vec{i} - 9{,}075\vec{j} = 33{,}387\ \underline{/-15.77°}\ \text{in./sec}^2$

Force Analysis—Link 3

$\Sigma\vec{T}_A = 0;\ \vec{R}_{G3/A} \times (-m_3\vec{A}_{G3}) + (-\bar{I}_3\vec{\alpha}_3) + \vec{R}_{B/A} \times$

$\vec{F}_{43} = 0$

Where

$\vec{R}_{G3/A} = \frac{1}{2}\vec{c} = 3.49\vec{i} + 1.97\vec{j};\ \vec{R}_{B/A} = \vec{c}$

$-m_3\vec{A}_{G3} = -1/12\left(\frac{0.7}{32.2}\right)(84{,}131\vec{i} - 39{,}052\vec{j}) =$

$-152{,}57\vec{i} + 70.82\vec{j} = 168.2\ \underline{/-24.9°}\ \text{lb}$

$-\bar{I}_3\vec{\alpha}_3 = -6{,}989\ (0.02)\ \vec{k} = -139.78\vec{k}\ \text{in.-lb}$

$\vec{F}'_{43} = \vec{F}_{43}\ \underline{/116.18°}\ \text{lb}$ Only direction is known.

Then

$$\vec{R}_{G3/A} \times (-m_3\vec{A}_{G3}) = \begin{vmatrix} \vec{i} & \vec{j} & \vec{k} \\ 3.49 & 1.97 & 0 \\ -152.6 & 70.82 & 0 \end{vmatrix}$$

$= 547.7\vec{k}\ \text{in.-lb}$

$$\vec{R}_{B/A} \times \vec{F}_{43} = \begin{vmatrix} \vec{i} & \vec{j} & \vec{k} \\ 6.98 & 3.94 & 0 \\ F'^{x}_{43} & F'^{y}_{43} & 0 \end{vmatrix}$$

$= (6.98\ F'^{y}_{43} - 3.94\ F'^{x}_{43})\ \vec{k}\ \text{in.-lb}$

Also

$\Sigma\vec{T}_A = 0$

$547.72\vec{k} - 139.78\vec{k} + (6.98\ F'^{y}_{43} - 3.94\ F'^{x}_{43})\ \vec{k} = 0$

$F'^{y}_{43}/F'^{x}_{43} = \tan 116.18° = -2.03$

Solving

$F'^{x}_{43} = 22.53\ \text{lb};\ F'^{y}_{43} = -45.73\ \text{lb}$

$\therefore F'_{43} = 22.53\vec{i} - 45.73\vec{j} = 50.98\ \underline{/-63.77°}\ \text{lb}$

Taking $\Sigma\vec{F}_3 = 0$ yields

$\vec{F}'_{43} + (-m_3\vec{A}_{G3}) + \vec{F}'_{23} = 0$

Or

$22.53\vec{i} - 45.73\vec{j} - 152.52\vec{i} + 70.82\vec{j} + \vec{F}'_{23} = 0$

$\vec{F}'_{23} = 129.99\vec{i} - 25.09\vec{j} = 132.39\ \underline{/-10.92°}\ \text{lb}$

Link 4

$\Sigma\vec{T}_{04} = 0;\ \vec{R}_{G4} \times (-m_4\vec{A}_{G4}) + (-\bar{I}_4\vec{\alpha}_4) +$

$\vec{R}_B \times \vec{F}_{34} = 0$

$\vec{R}_{G4} = 1/2\ \vec{d} = -1.32\vec{i} + 2.7\vec{j} = 3\ \underline{/116.18°}\ \text{in.}$

$\vec{R}_B = \vec{d} = -2.65\vec{i} + 5.39\vec{j} = 6\ \underline{/116.18°}\ \text{in.}$

$m_4\vec{A}_{G4} = -1/12\left(\frac{0.8}{32.2}\right)(32{,}130\vec{i} - 9{,}075\vec{j}) =$

$-66.52\vec{i} + 18.79\vec{j} = 69.12\ \underline{/164.23°}\ \text{lb}$

$-\bar{I}_4\vec{\alpha}_4 = (0.01)\ (-8{,}268.4)\ \vec{k} = 82.68\vec{k}\ \text{in.-lb}$

$\vec{F}''_{34} = F_{34}\ \underline{/29.44°}\ \text{lb}$ Only direction is known.

Then

$\vec{R}_{G4} \times (-m_4\vec{A}_{G4}) = 154.80\vec{k}$

$\vec{R}_B \times \vec{F}''_{34} = (-2.65\ F''^{y}_{34} - 5.39\ F''^{x}_{34})\vec{k}$

And

$\Sigma\vec{T}_{04} = 0$

$$154.80\vec{k} + 82.68\vec{k} + (-2.65\, F_{34}''^{y} - 5.39\, F_{34}''^{x})\,\vec{k} = 0$$

$$F_{34}''^{y}/F_{34}''^{x} = \tan 29.44° = .564$$

$$\therefore\ \vec{F}_{34}'' = 39.72\vec{i} + 22.40\vec{j} = 45.60\ \underline{/29.42°}\ \text{lb}$$

Also

$$\Sigma\vec{F}_4 = 0 \text{ yields}$$

$$\vec{F}_{34}'' + (-m_4\vec{A}_{G4}) + \vec{F}_{14}'' = 0$$

Or

$$39.72\vec{i} + 22.40\vec{j} - 66.52\vec{i} + 18.79\vec{j} + \vec{F}_{14}'' = 0$$

$$\vec{F}_{14}'' = 26.80\vec{i} - 41.19\vec{j} = 49.14\ \underline{/-56.95°}\ \text{lb}$$

$$\vec{F}_{32} = \vec{F}_{32}' + \vec{F}_{32}'' = -\vec{F}_{23}' - \vec{F}_{23}'' = -129.99\vec{i} + 25.09\vec{j} - 39.72\vec{i} - 22.40\vec{j} = -169.71\vec{i} + 2.69\vec{j} = 169.73\ \underline{/179.09°}\ \text{lb}$$

Link 2

$$\Sigma\vec{T}_{02} = 0;\ \vec{R}_A \times \vec{F}_{32} + \vec{T}_2 = 0$$

$$\vec{T}_2 = -\vec{R}_A \times \vec{F}_{32} = (-2.6\vec{i} + 1.5\vec{j}) \times (-169.71\vec{i} + 2.69\vec{j})$$

$$\vec{T}_2 = 248\vec{k}\ \text{in.-lb ANS.}$$

Note:

1. The difference in answer when compared with graphical solution is due to drawing error.
2. If $m_4 = 0$; $\vec{T}_2 = (-2.6\vec{i} + 1.5\vec{j}) \times (-129.99\vec{i} + 25.09\vec{j}) - 65.23\vec{k} + 194.99\vec{k} =$
   $T_2 \simeq 129.76k$ in.-lb. which is practically the same as in Table 2-2.

Force Analysis Using Complex Numbers Method

In order to extend the TI-59 program to force analysis, the following data must be added to the kinematic part:

1. The linear acceleration of the center of gravity of the links
2. The mass and moment of inertia (m and $\bar{I}$ for each link)
3. The equilibrium equations, in order to determine the forces and torques

The following procedures use previously calculated data.

Link 3 (See Figure 2-21).

$$r = r_{g3} + \frac{h_3}{\sin(\theta_{AG_3} - \theta_3)} = 4 + \frac{.82}{\sin(-26.43 - 29.19)}$$

$$= 3.01$$

Writing $\Sigma\vec{F} = 0$ for link 3 yields:

$$\vec{F}_{23} + \vec{F}_{43} + m_3\vec{A}_{G3} = 0$$

or when using complex numbers:

$$F_{23}e^{j\theta_{F23}} + F_{43}e^{j\theta_4} + m_3A_{G3}e^{j(\theta_{AG3} + 180°)} = 0$$

The Equation can be rewritten as an equivalent system of two equations by separating the real and imaginary parts or

$$F_{23}\cos\theta_{F_{23}} + F_{43}\cos\theta_4 + m_3A_{G3}\cos(\theta_{A_{G3}} + 180°) = 0$$

$$F_{23}\sin\theta_{F_{23}} + F_{43}\sin\theta_4 + m_3A_{G3}\sin(\theta_{A_{G3}} + 180°) = 0$$

Plus the $\Sigma\vec{M}_A = 0$ equation which is

$$F_{43}c\sin(\theta_4 - \theta_3) - m_3A_{G3}r\sin(\theta_{A_{G3}} - \theta_3) = 0$$

This will give you a system of three equations with the three unknowns, i.e., magnitudes of F_{43}, F_{23} and the direction θ_{F23}.

The solution follows:

$$F_{43} = (168.5)\frac{(3.01)(-.8252)}{8\sin 85.82} \approx -52.45\ \text{lb}$$

Real→ $F_{23}\cos\theta_{F23} = -52.45\cos 115° - 168.5\cos(-26.43 + 180°) = 128.72$

Imaginary → $F_{23}\sin\theta_{F23} = -52.45\sin 115° - 168.5\sin(153.57°) = -27.5$

$$\text{Tan}\,\theta_{F_{23}} = \frac{-27.5}{128.7} \rightarrow \theta_{F23} = -12.06°$$

$$F_{23} = -131.62\ \text{lb}\ \underline{/-12.06°}$$

$$F_{14} = -F_{43}$$

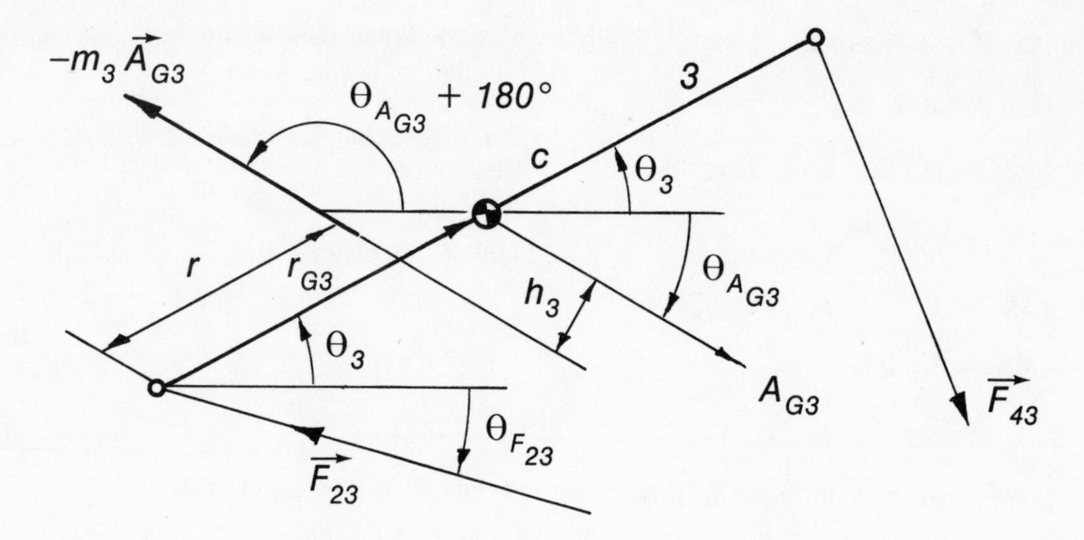

Figure 2-21. Inertia force analysis of link 3.

Considering the $\Sigma\vec{M}_{02} = 0$ for link 2, gives the torque T_2' (due to m_3A_{G3}) as:

$$T_2' = -F_{23}b \sin(\theta_2 - \theta_{F_{23}})$$

$$T_2' = (-131.62)\ 3 \sin(150° + 12.06°) =$$

$$-121.36 \text{ in.-lb}$$

Link 4 (See Figure 2-22)

$$r_4 = r_{G4} + \frac{h_4}{\sin(\theta_{A_{G4}} + 180 - \theta_4)} =$$

$$3 + \frac{1.20}{\sin(164 - 115)} = 4.59 \text{ in.}$$

$$\Sigma\vec{F} = 0$$

$$\vec{F}_{14} + \vec{F}_{34} + m_4\vec{A}_{G4} = 0$$

Or

$$F_{14}\cos\theta_{F_{14}} + F_{34}\cos\theta_3 + m_4A_{G4}\cos(\theta_{A_{G4}} + 180°) = 0$$

$$F_{14}\sin\theta_{F_{14}} + F_{34}\sin\theta_3 + m_4A_{G4}\sin(\theta_{A_{G4}} + 180°) = 0$$

and

$$\Sigma\vec{M}_{0_4} = 0$$

$$F_{34}d \sin(\theta_4 - \theta_3) + m_4A_{G4}r_4 \sin(\theta_{A_{G4}} - \theta_4) = 0$$

Then,

$$F_{34} = 69.4\,\frac{4.59 \sin(-15.77° - 115°)}{6 \sin(115° - 29.18°)} = -40.31 \text{ lb}$$

$$F_{14}\cos\theta_{F_{14}} = -(69.4)(-.9613) - (40.31)(.8731) =$$

$$31.51 \text{ lb}$$

$$F_{14}\sin\theta_{F_{14}} = (-69.4)(.2756) - (40.31)(.4875) =$$

$$-38.77 \text{ lb}$$

$$\therefore F_{14} = 49.96 \ \underline{/-50.90°} \text{ lb}$$

$$T_2'' = F_{34}b \sin(\theta_2 - \theta_3) = (40.32)(3) \sin 120 =$$

$$104.75 \text{ in.-lb}$$

$$\text{Total Torque } T = T' + T'' = 226.11 \text{ in.-lb}$$

Shaking Forces and Moments

Vibration and inertia effects are often transmitted to foundations and surroundings. They are called "shaking forces." In general, the shaking force is the vector sum of the variable forces acting on the frame of machines due to static forces and inertia forces. They are separated because sometimes the inertia forces can be balanced, and quite often the static shaking forces are negligible in comparison with the inertia shaking forces. In the problem, assume such to be the case and neglect the static shaking force effect.

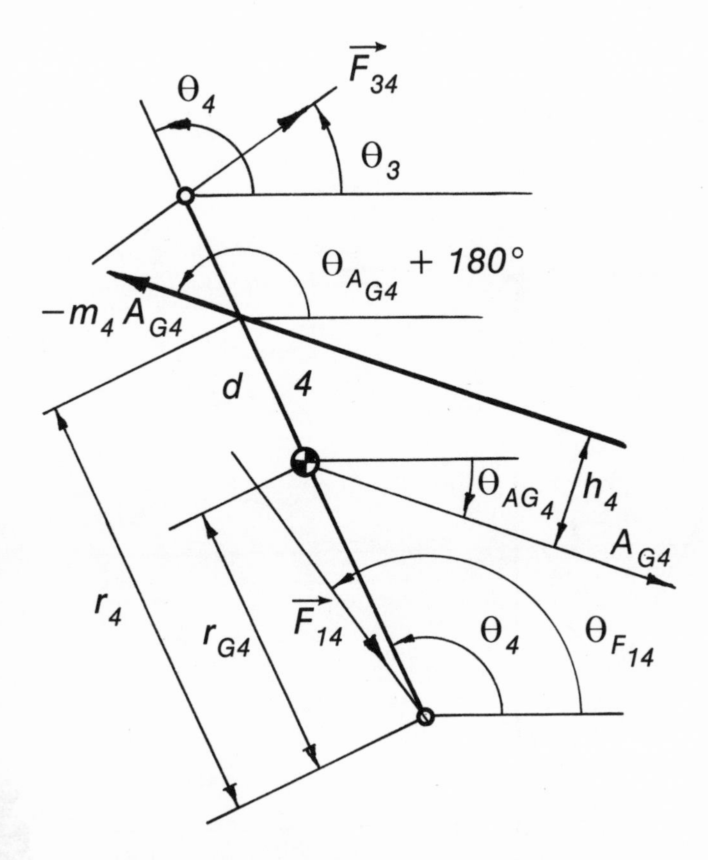

Figure 2-22. *Inertia force analysis of link 4.*

By definition, the shaking force due to inertia is the vector sum of the inertia forces of the individual links or $\vec{F}_s = -\sum_{i=1}^{n} (m_i\vec{A}_{Gi})$. Such shaking forces may interfere with the operations of other machines and instruments. The inertia force could be quite large and vary widely in magnitude, producing an unwanted fatigue effect on the operation of the machine. There are methods for balancing the shaking forces, and they should be used whenever possible. To acquire a better understanding of shaking forces, a "Flying Shear Mechanism" problem is given. The shaking forces will be $m\vec{a}_3$ and $m\vec{a}_4$. You will be concerned with the line of action of the shaking force with respect to the positions of the machine because this will influence the design of the foundation and mounts.

You should also examine the sum of the inertia torques $(I_i\vec{\alpha}_i)$ and moments due to intertia torques $(\vec{r}_i \times m_i\vec{A}_{Gi})$ about a point on the frame. The summation of these is called the "shaking moment." Additionally, consider the sum of all torques due to external loading. As with the shaking forces, for a constant input speed, the shaking moment will vary as the crank moves cyclically.

Flying Shear Mechanism[3]

The main objective of this example is to show how the shaking forces are determined. Normally, an additional static force analysis including the addition of the cutting force at point C (shown in Figure 2-23) would be required. This cutting force could be found from the characteristics of the materials being cut, their size, etc.; e.g., 500 lb. Assume that $\omega_2 = 50$ rad/sec CCW (Figure 2-23); links are made of steel plates 1 in. thick (specific weight 490 lb/ft³); and that the moments of inertia about the centroidal axis $\bar{I}_4 = 25$ ft-lb-sec² ($m_4 = 2.66$ lb-ft-sec² or slugs). Using a $K_S - 1'' = 1'$ scale, the dimensions needed to work this problem would be outlined. When using a graphical approach (for checking the program and other quick needs) the recommended scales for velocity and acceleration polygons are given in Figure 2-27.

Two positions of the crank are shown in the solutions and the shaking forces drawn to scale. Table 2-3 contains the results obtained by use of a program similar to the one previously described and included in Appendix 1. In order to save space, parts of the required step-by-step solutions were omitted and some other short cuts made. For example, the center of gravity of link 3 being close to the centerline of the AB portion of link 3 was considered to be on this line at a distance of 2.2 feet from point A. To obtain the results of Table 2-3, nine values must be stored in the program (i.e., a, b, c, d, ω_2, θ_2, W_3, I_3 and r_{G3}). Next, a short program for calculating the inertia effects has been added and executed. Additionally, an outline of the procedure for finding the moment of inertia for link 3 has been shown. Assuming that link 4 would have a triangular shape, the moment of inertia could be found in a manner similar to the example of the rock crusher which also has triangular shape links. But, the best way is to find the moments of inertia experimentally by measuring the frequency of oscillation of the links.

Assume that the following results were obtained (and used) in the example:

$\bar{I}_3 = 6.38$ lb-ft-sec²
$\bar{I}_4 = 25.00$ lb-ft-sec²
$m_3 = 1.83$ lb-sec²/ft (or slugs)
$m_4 = 2.66$ slugs

In order to show how moments of inertia could be obtained analytically, calculations are included on the following pages. It is expected that the result will differ from those obtained experimentally.

The HP-67 pocket calculator has a preprogrammed set of formulas to determine moments of inertia of rectangular and circular combinations of shapes.

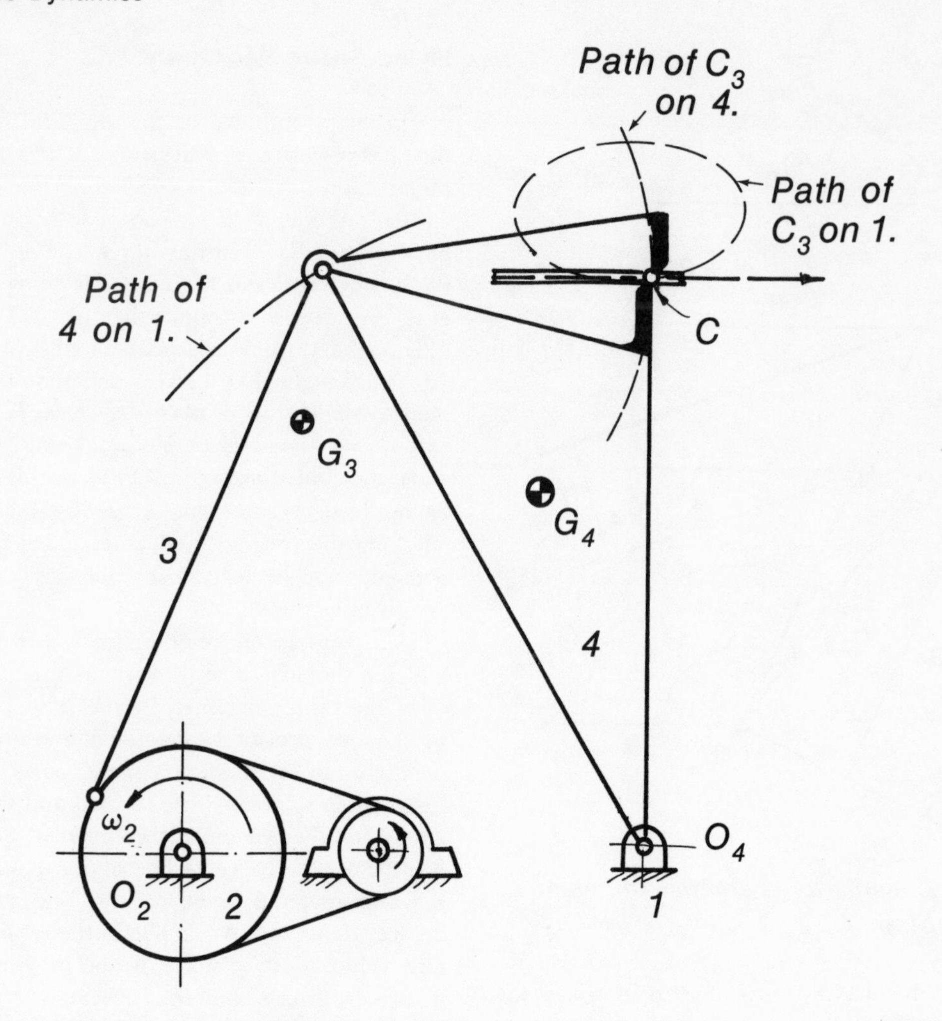

Figure 2-23. *Flying shear mechanism.*

The answer for this example, when checked by use of the package program 02-A2, was $I_{zz} = 3.115$ ft^4 which is practically the same as the graphical and analytical solutions. Figure 2-24 shows the input values and results for a slightly modified shape for this link, and Figure 2-26 shows a similar solution for link 4.

You can determine the moment of inertia I_{zz} simply by using the following formulas:

$$\bar{I}_{zz} = \bar{I}_{xx} + \bar{I}_{yy} = \bar{I}_u + \bar{I}_v$$

and

$$I_{zz} = \sum_{i=1}^{3} \left(\frac{b_i h_i^3}{12} + A_i d_i^2 \right)$$

Substituting dimensions yields:

$$\bar{I}_{zz} = \frac{(0.3)\ (3)^3}{12} + (0.3)\ (3)\ (0.9)^2 + \frac{(1.86)\ (0.3)^3}{12} +$$

$$(.3)\ (1.86)\ (0.7)^2 + 0.00127 + 0.05754 +$$

$$\frac{(3)\ (0.3)^3}{12} + (0.3)\ (3)\ (0.4)^2 + \frac{(0.3)\ (1.86)^3}{12} +$$

$$(0.3)\ (1.86)\ (1.05)^2 + 0.000833 + 0.4264 =$$

$$3.0947 \text{ ft}^4 \text{ ANS. (See Figure 2-25.)}$$

Note: The axis are inclined, except for the last element which was copied directly from Table 2-4, part 15. Of course, this method is much simpler when compared with the graphical solution. The graphical solution in Figure 2-25 is shown to make you aware of it in case

Table 2-3
Shaking Force Analysis

θ_2	ω_3	ω_4	α_3	α_4	m_3A_{g3}	m_4A_{G4}	h_3	h_4
0	See 360° For Results							
30	−15.65	−10.03	229.62	672.64	3,869.15	3,618.24	.38	4.65
60	−10.89	− 2.08	592.22	754.25	4,085.11	4,012.68	.92	4.7
90	− 4.55	4.62	581.47	503.12	2,703.66	2,679.06	1.38	4.7
120	1.07	8.41	484.74	228.82	1,425.89	1,274.14	2.17	4.5
150	5.60	9.64	380.48	18.61	1,049.85	504.20	2.31	.92
180	9.02	90.2	268.58	−129.50	1,389.83	813.63	1.23	4.0
210	11.11	7.07	124.34	−237.53	1,812.93	1.291.34	.44	4.6
240	11.42	4.08	− 78.72	−334.65	2,222.48	1,782.54	.23	4.7
270	9.17	− .002	−366.87	−449.22	2,634.14	2,389.85	.89	4.7
300	3.45	− 5.36	−728.76	−566.73	2,874.35	3,018.88	1.62	4.7
330	− 5.64	−11.27	−945.58	−502.56	2,122.57	2,757.68	2.84	4.5
360	−14.10	−14.10	−545.53	42.74	973.69	1,081.83	3.57	.99

Table 2-4
Moment of Inertia by Graphical Method

No.	Area	$\bar{x}$	$\bar{y}$	$\bar{x}A$	$\bar{y}A$	d_x	d_y	Ad^2_x	Ad^2_y	$\bar{I}_{xx}$ 10^{-3}	$\bar{I}_{yy}$ 10^{-3}
1	.017	.21	.10	.0036	.0017	1.27	2.05	.0277	.0722	.0149	0.722
2	.128	.27	.33	.0345	.0421	1.21	1.81	.1869	.4182	1.4600	1.2700
3	.128	.42	.69	.0536	.0881	1.05	1.45	.1407	.2684	1.4600	1.2700
4	.128	.59	1.05	.0753	.1340	.89	1.06	.1011	.1434	1.4600	1.2700
5	.128	.75	1.42	.0957	.1813	.72	.70	.0662	.0625	1.4600	1.2700
6	.128	.91	1.81	.1162	.2311	.57	.32	.0415	.0131	1.4600	1.2700
7	.128	1.10	2.17	.1404	.2770	.38	.06	.0184	.0005	1.4600	1.2700
8	.128	1.27	2.53	.1621	.3229	.23	.42	.0067	.0225	1.4600	1.2700
9	.017	1.32	2.75	.0227	.0472	.19	.68	.0006	.0079	.0149	.0722
10	.035	1.68	2.83	.0592	.0997	.17	.70	.0010	.0173	.0349	.0546
11	.128	1.93	2.92	.2470	.3738	.43	.81	.0237	.0840	.1980	9.83
12	.186	2.34	3.08	.4357	.5735	.78	.95	.1133	.1681	.3040	27.40
13	.129	2.81	3.21	.3633	.4151	1.38	1.08	.2462	.1508	.1910	10.10
14	.042	3.08	3.32	.1294	.1394	1.60	1.20	.1075	.0605	.0415	.9260
15	.111	3.46	2.93	.3841	.3252	1.96	.72	.4264	.0575	1.2700	.8330
Σ	1.5610							1.5081	1.5469	.0123	.0587

an irregular (which is the case for most real links) shape needs to be worked.

The calculation of the moment of inertia I_{zz} is shown twice to help you get a good understanding of this minor issue. Past experience illustrates that it is not so much the overall concept of solving the problem which causes the most mistakes, but rather little things such as units, dimensions, geometry, etc.

Then,

$$\bar{I}_{xx} = 1.5469 + 0.0123 = 1.5592 \text{ ft}^4$$

$$\bar{I}_{yy} = 1.5081 + 0.0587 = 1.5668 \text{ ft}^4$$

$$\bar{I}_{zz} = \bar{I}_{xx} + \bar{I}_{yy} = 3.1259 \text{ ft}^4$$

$$\bar{I}_{m,zz} = (3.1259)\ (490)/(386) = 3.964 \text{ ft-lb-sec}^2$$

$$\text{Mass } m_3 = (1.5610)\ (490)/(386) = 1.98 \text{ slugs}$$

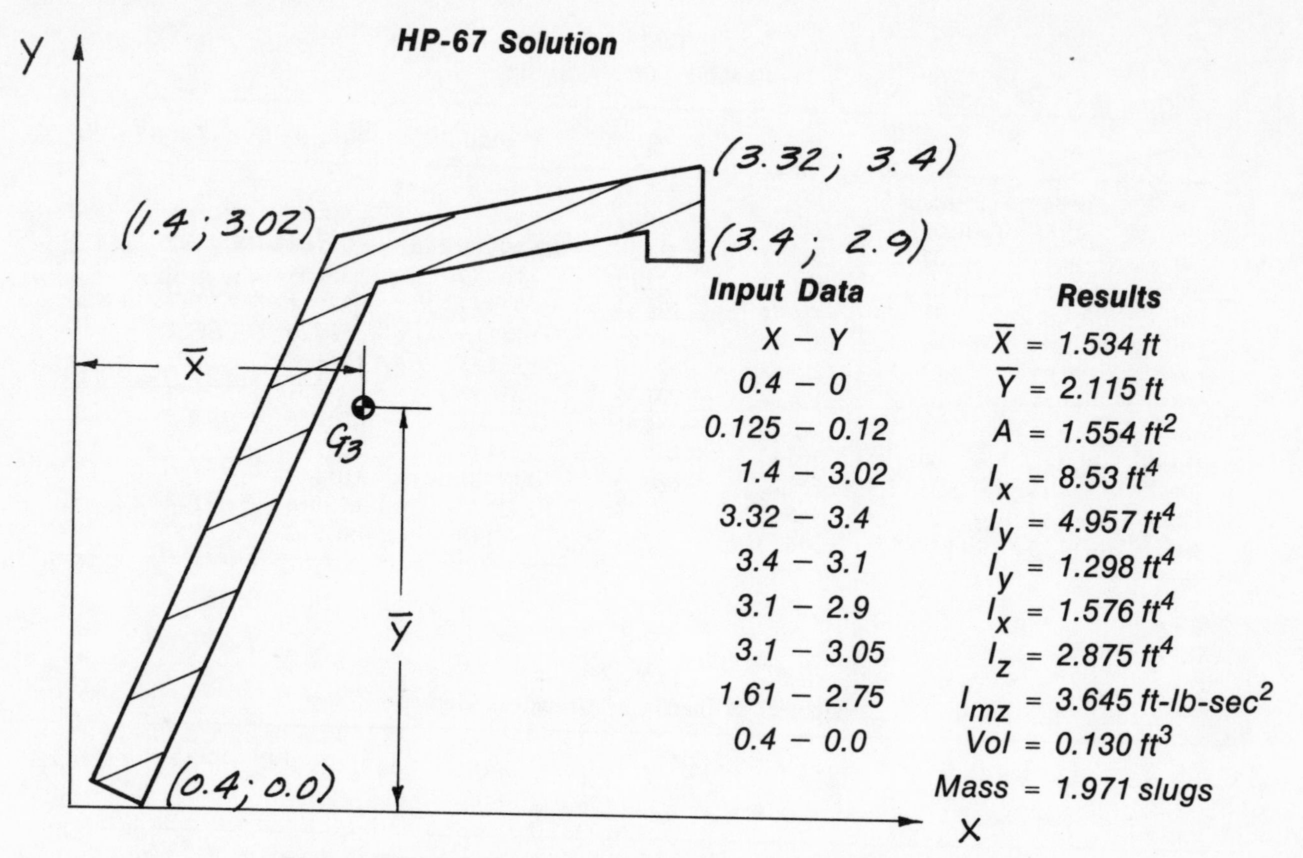

Figure 2-24. *Moment of inertia of link 3 by analytical method.*

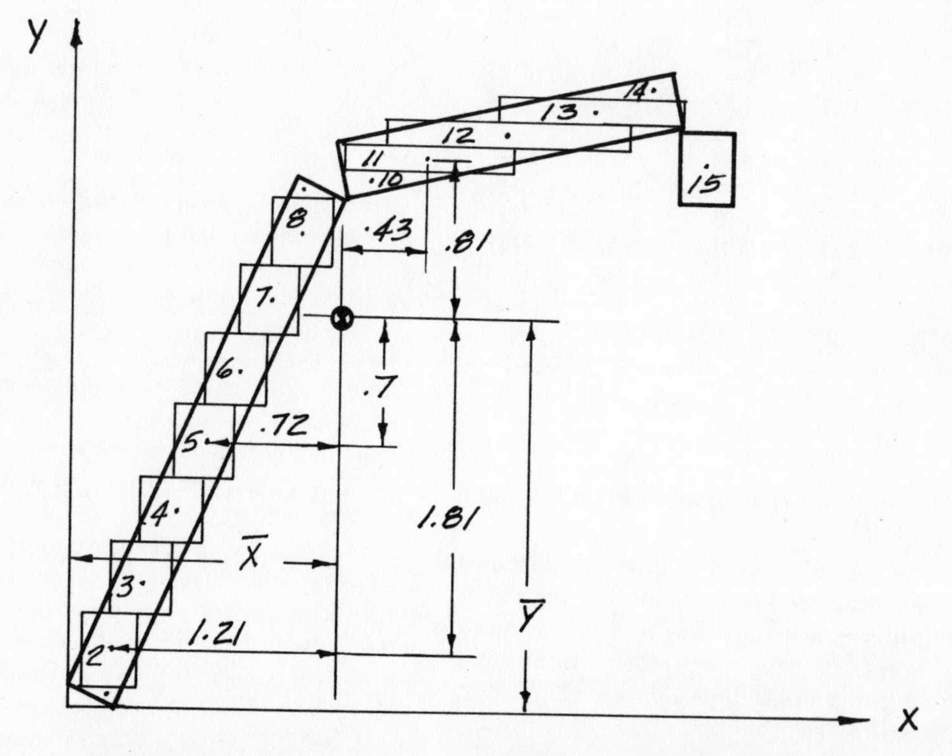

Figure 2-25. *Moment of inertia of link 3 by graphical solution.*

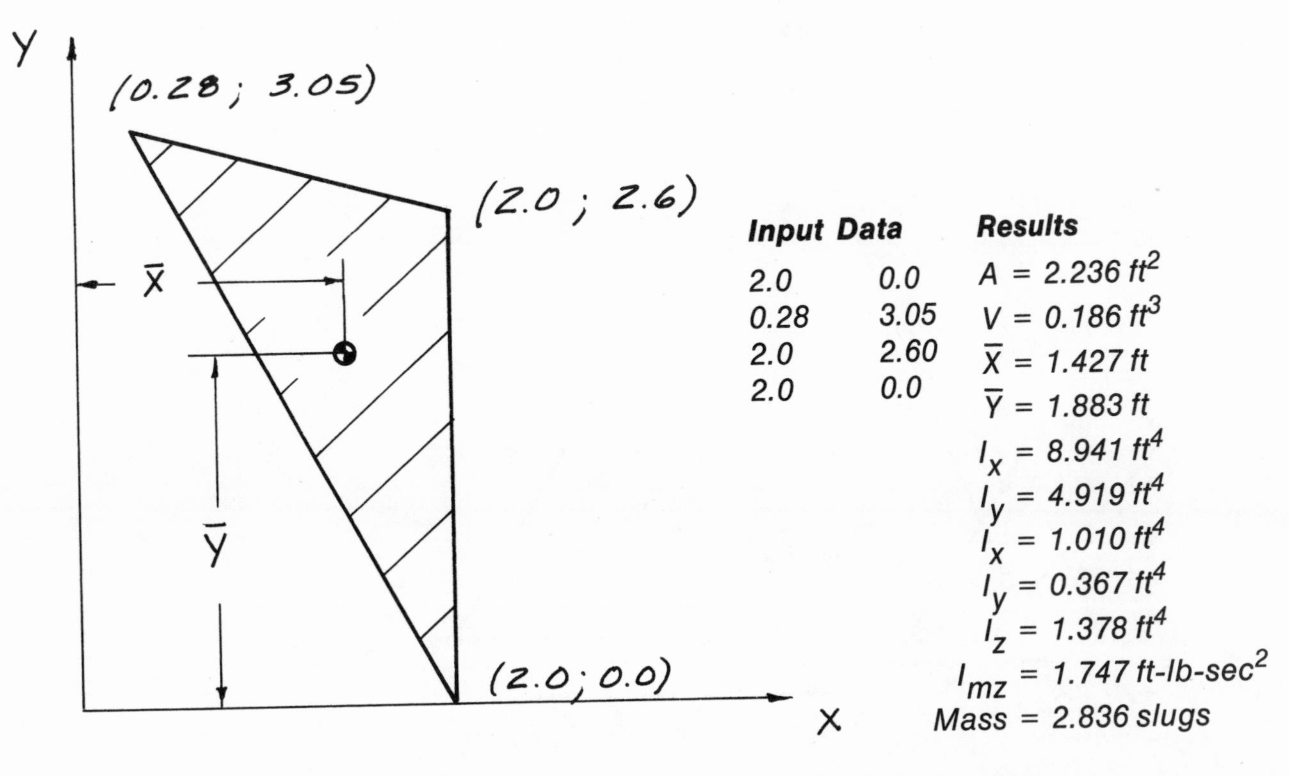

Figure 2-26. *Moment of inertia of link 4 by analytical method.*

In order to show how the table has been obtained, see Figure 2-25 and the following examples:

$$A_1 = 1/2bh = (.5)\ (.275)\ (.125) = 0.017\ \text{ft}^2$$

$$A_2 = A_8 = bh = (.345)\ (.37) = 0.128\ \text{ft}^2$$

$$\bar{x} = 1.489;\ \bar{y} = 2.085$$

$$\bar{I}_{xx,1} = \frac{1}{36}bh^3 = \left(\frac{1}{36}\right)(.275)\ (.125)^3 = (.0149)\ (10^{-3})\ \text{ft}^4$$

$$\bar{I}_{xx,2} = \frac{1}{12}bh^3 = \left(\frac{1}{12}\right)(.345)\ (.37)^3 = 1.46\ (10^{-3})\ \text{ft}^4$$

$$\bar{I}_{yy,1} = \frac{1}{36}b^3h = \left(\frac{1}{36}\right)(.275)^3\ (.125) = (.0722)\ (10^{-3})\ \text{ft}^4$$

$$\bar{I}_{yy,2} = \frac{1}{12}b^3h = \left(\frac{1}{12}\right)(.345)^3\ (.37) = 1.27\ (10^{-3})\ \text{ft}^4 \text{ etc.}$$

Comments

For the two inertia forces, $m_3\vec{A}_{g3}$ and $m_4\vec{A}_{g4}$, the resultant shaking force is obtained by vectorially adding the two forces. The shaking force, $\vec{S}$, will act through the point of intersection of the two lines of action of the inertia forces. (See Figures 2-27 and 2-28.)

The magnitude and the direction of the shaking force $\vec{S}$ varies with the rotation of the crank.

In this example, you may observe that:

1. For the $\theta_2 = 150°$ position, the inertia effect of link 4 is minimal (504 lb), but the shaking force (approx. 1,300 lb) passes through the bearing at point 0_2 (Figure 2-27)
2. For the $\theta_2 = 60°$ position, both inertia forces (4,085 and 4,012 lb) are maximum, but the line of action of the shaking force (equal to approx. 8,000 lb) lies outside the machine mounting (Figure 2-28)

Again, with the use of a programmable calculator, you can quickly run through the whole cycle and determine the magnitudes of the inertia forces. Of course, to obtain a complete picture, a more detailed study must be made.

Other graphical and analytical means of handling this task (e.g., polar shaking force curves) are described in the literature. The objective of this section is to study how inertia shaking forces affect the design, and how they are determined for such a machine.

In Chapter 3, methods of balancing are briefly discussed. Some of them will apply to alleviating the unwanted effects of the shaking forces just determined.

(Text continued on page 40)

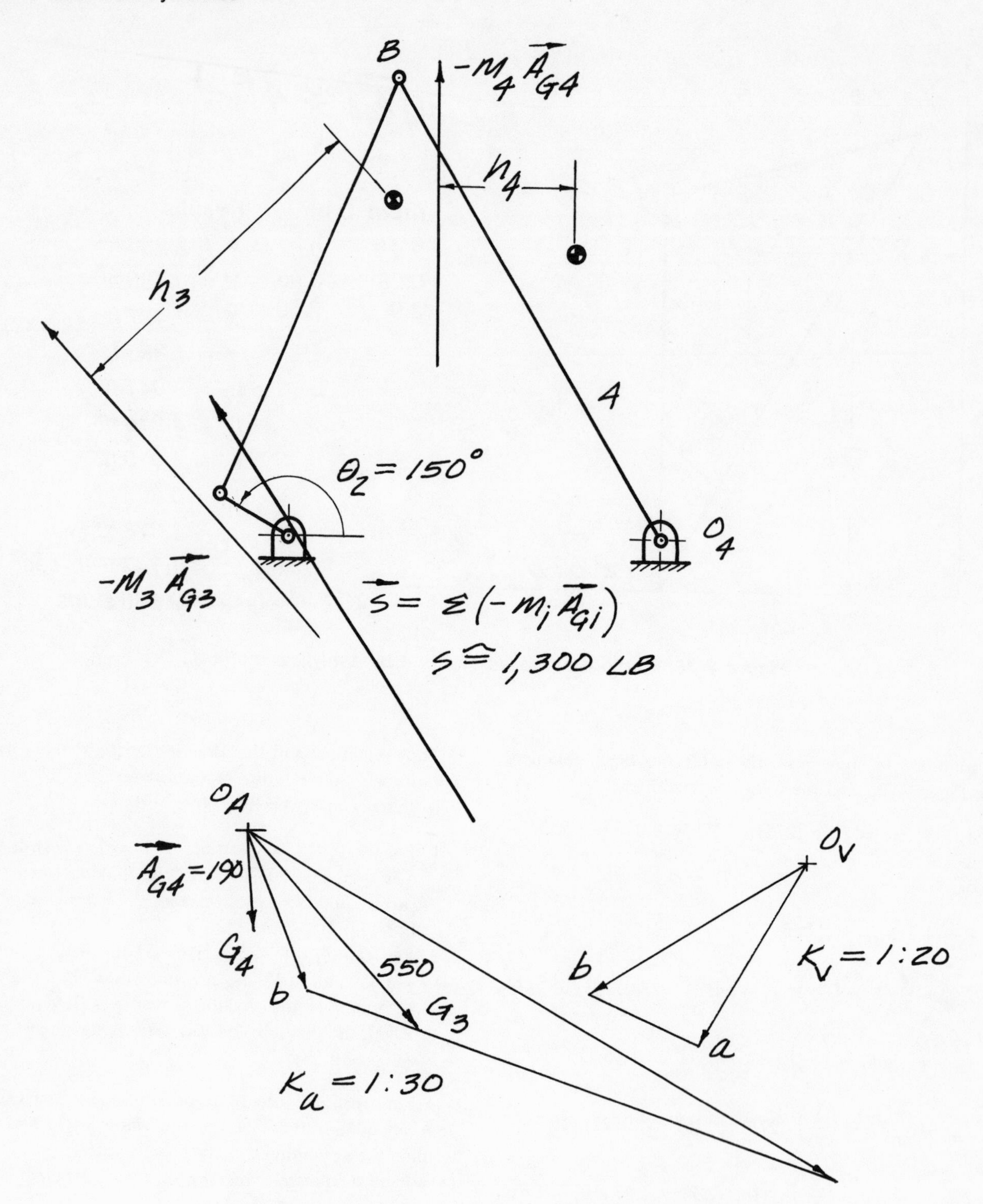

Figure 2-27. *Shaking force for flying shear mechanism.*

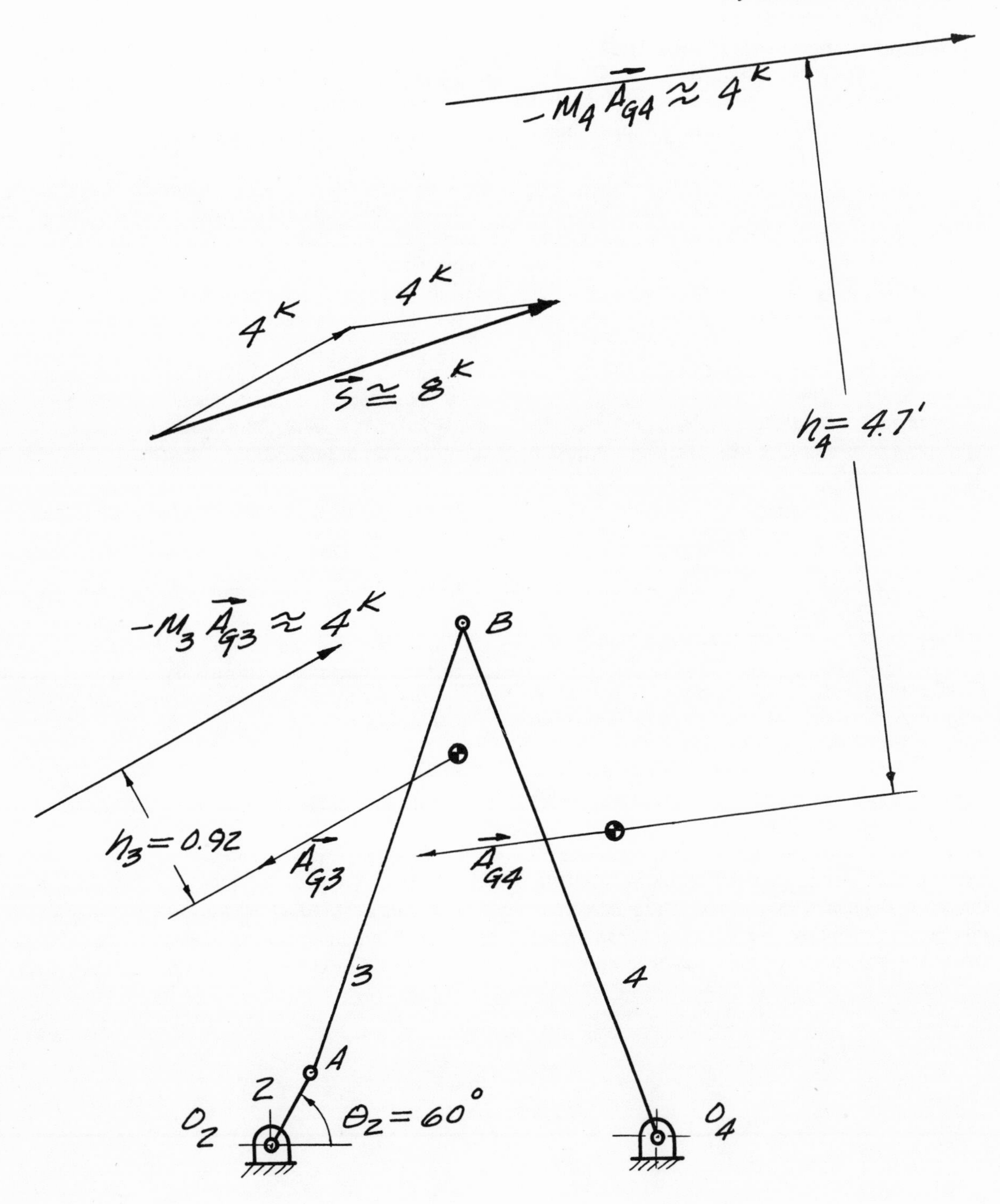

Figure 2-28. *Shaking forces for flying shear mechanism.*

Balancing the Shaking Force of a Four-Bar Linkage

The four-bar flying shear mechanism will be balanced by using the following steps:

1. If the center of gravity of link 3 is away from the line of pin centers A and B, add a mass on the opposite side of the coupler's center line, bringing the combined mass center to the coupler line. In this case, it is assumed for simplicity that the coupler's center of gravity is on the coupler center line. Therefore, this step was omitted.

   Proportionally divide the coupler mass into two equivalent masses to be applied at points A and B in order to satisfy the two equations:

   $$m_A r_{G3} = m_B (c - r_{G3})$$

   $$m_A + m_B = m_3$$

   Assume for this mechanism, $c = 3$ ft; $r_{G3} = 2.2$ ft; and $m_3 = 1.83$. Solving the above equations yields, $m_A = 0.49$ and $m_B = 1.34$.
2. Balance the crank with the equivalent portion of m_A due to the weight of link 3. (Note, the weight of the crank is assumed to be zero in the problem.) This will relocate the center of gravity of the crank to the crankpin O_2. (See Figure 3-1 in Chapter 3.)
3. By adding the correction weight, also balance the driven link 4, considering its weight located at a distance of 2 feet as shown, and also considering the newly calculated weight, m_B. The total correction weight could be placed at any convenient radius. This is shown in Figure 2-29 to equal 3.5 feet, and in Figure 2-30 to be 1 foot. When these steps are taken, the four-bar linkage will be force-balanced—the shaking force will always be zero for any position of the mechanism.

This could be seen best in Figure 2-29 and in Figure 2-30, where the inertia force vectors are equal in magnitude and opposite in sense. The drawings were repeated twice on purpose, in order to show the procedure and results without overcrowding.

The shaking moments will not be balanced since the angular accelerations of the driver and driven cranks are not zero. The linear accelerations at O_2 and O_4 are zero. Therefore, the inertia forces will also be zero for these links. It is very difficult to balance the shaking moments, if not impossible.

Some additional comments to be made at this point are:

1. Be cautious to avoid errors in calculating the moments of inertia which have changed due to the additions and the displacing of masses
2. The dynamic force analysis must be performed after all the masses are relocated and the linkage balanced
3. The mass of link 3 must be included in the analysis, since the weight of link 3 was not removed, and only two corrective measures were added to link 2 and link 4
4. Some other types of linkages (including a six-bar linkage) could be completely force balanced by the so-called "Method of Linearly Independent Vectors" and other techniques as described in literature on this subject.

(Text continued on page 43)

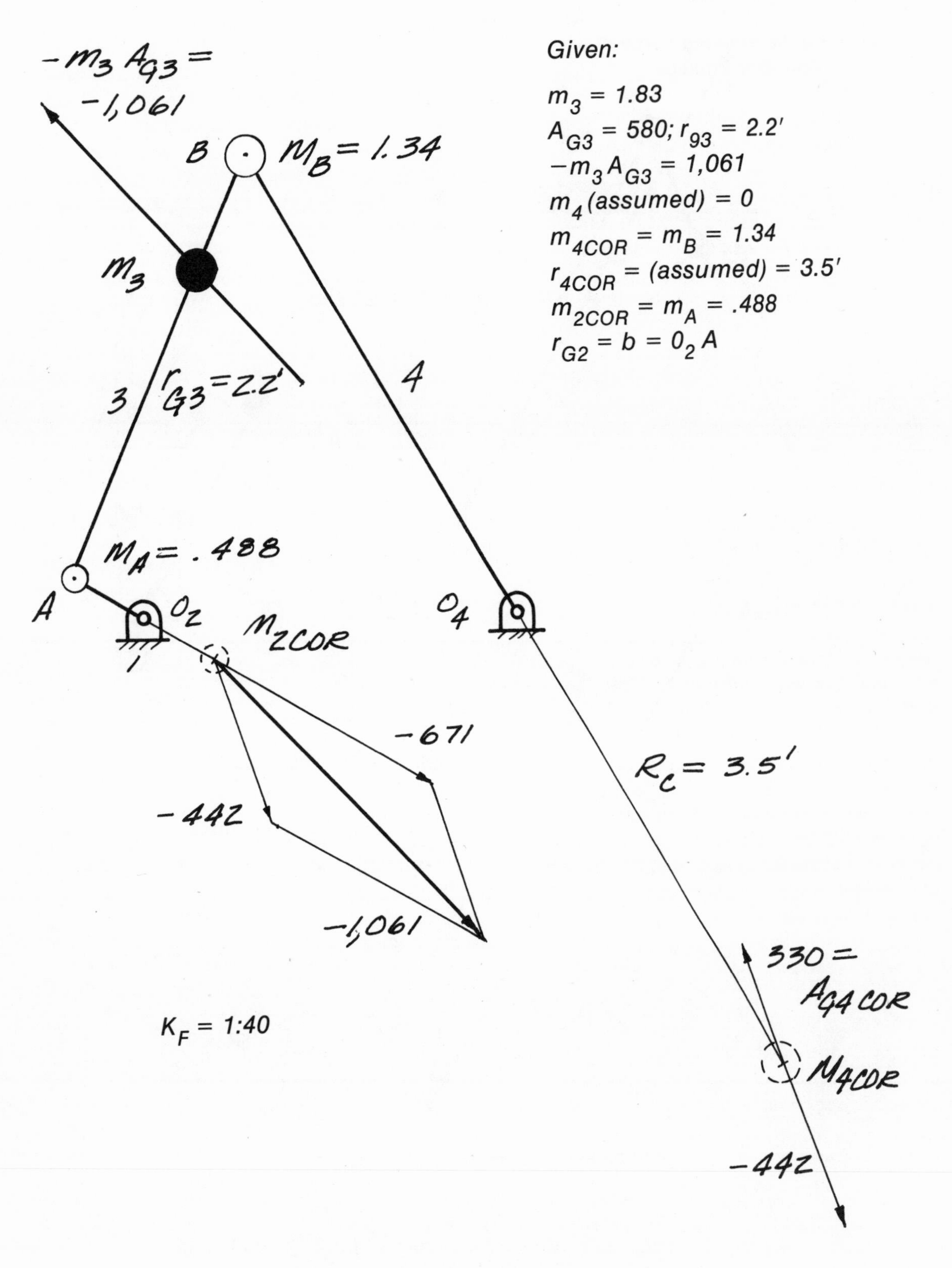

Figure 2-29. *Force-balancing (part 1).*

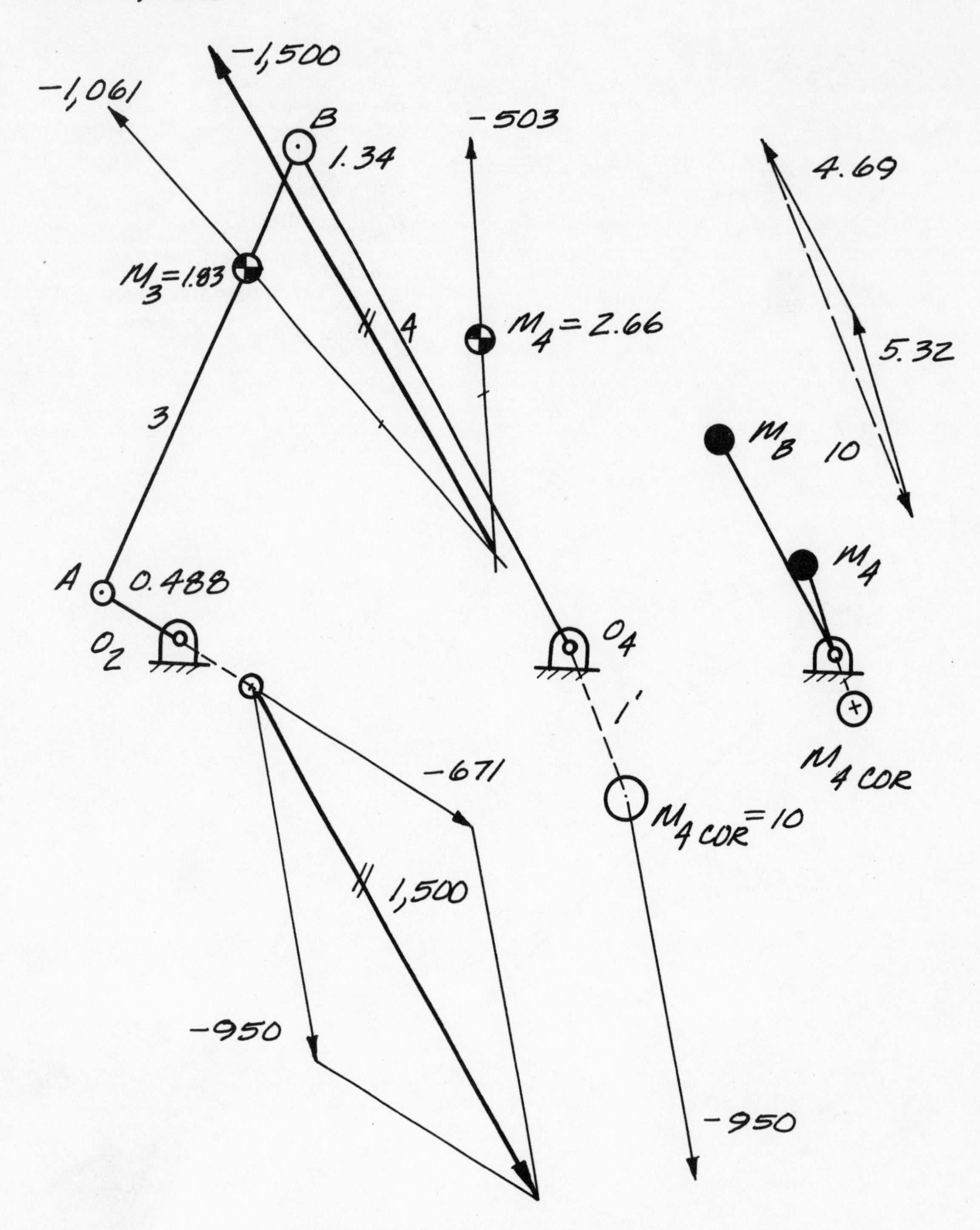

Figure 2-30. *Force-balancing (part 2).*

Rock Crusher

As another example, work the rock crusher problem sketched in Figures 2-31 and 2-32. This machine is a huge one, with heavy links 3 and 6. The driver is link 2, rotating with 572 rpm (rev/min) or 60 radians per second clockwise. Most machines operate more efficiently at high speeds. Therefore, the 60 rad/sec. rotation is probably reasonable for this machine. It would be rather difficult to conduct a complete static and dynamic force analysis of this machine. These calculations are limited to one position and the governing inertia effects of link 3 and 6 only are considered.

The crushing forces will be large due to the mass involved and the characteristics of the linkages. The limits of motion for link 4 (of the crank-and-rocker linkage on the right) and link 6 plus the approximate coupler curve for the connecting pin joint C should be examined. Notice that link 6 swings merely 5° of the 30° rocking angle for link 4 or the 360° of motion of crank 2. The analysis as shown in Figure 2-35 is not complete. You would have to consider the weights and other static forces for a complete solution. Also, for this particular position, the acceleration is not necessarily the maximum acceleration. Therefore, to shorten the solution, only the inertia effects have been accounted for, resulting in the forces acting as shown.

If a more complete solution is needed, the instructions given in solving the shovel mechanism would be helpful and should be sufficient to solve this problem. Also, a static force analysis for a similar problem involving a six-bar linkage is shown in the section "Direct Use of Newton's Laws."

(Note: When designing a rock crusher machine, you must conduct an in-depth study of the actual link proportions—link 3 and link 6 would most likely be switched.)

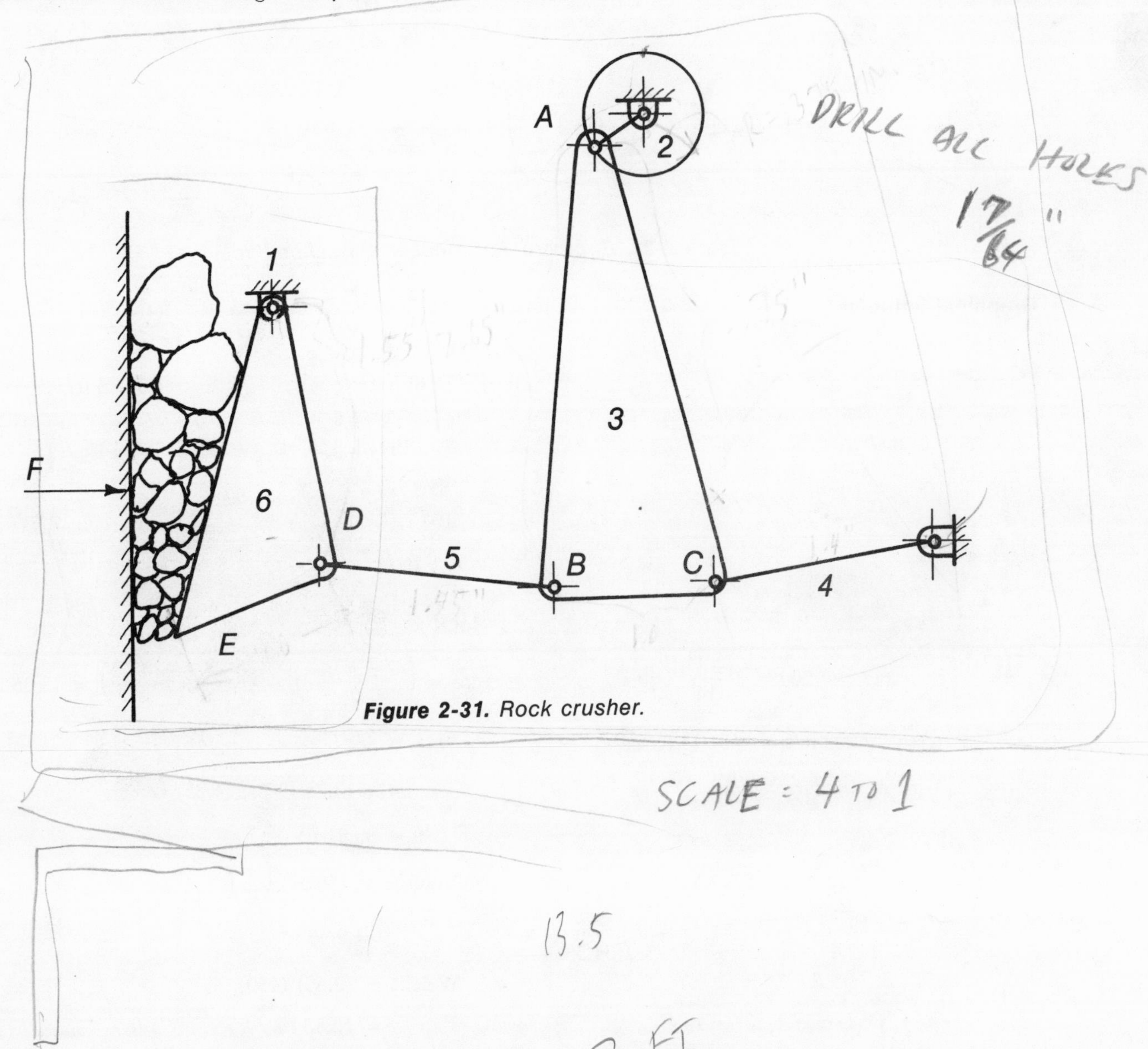

Figure 2-31. Rock crusher.

SCALE = 4 TO 1

13.5

2 FT.

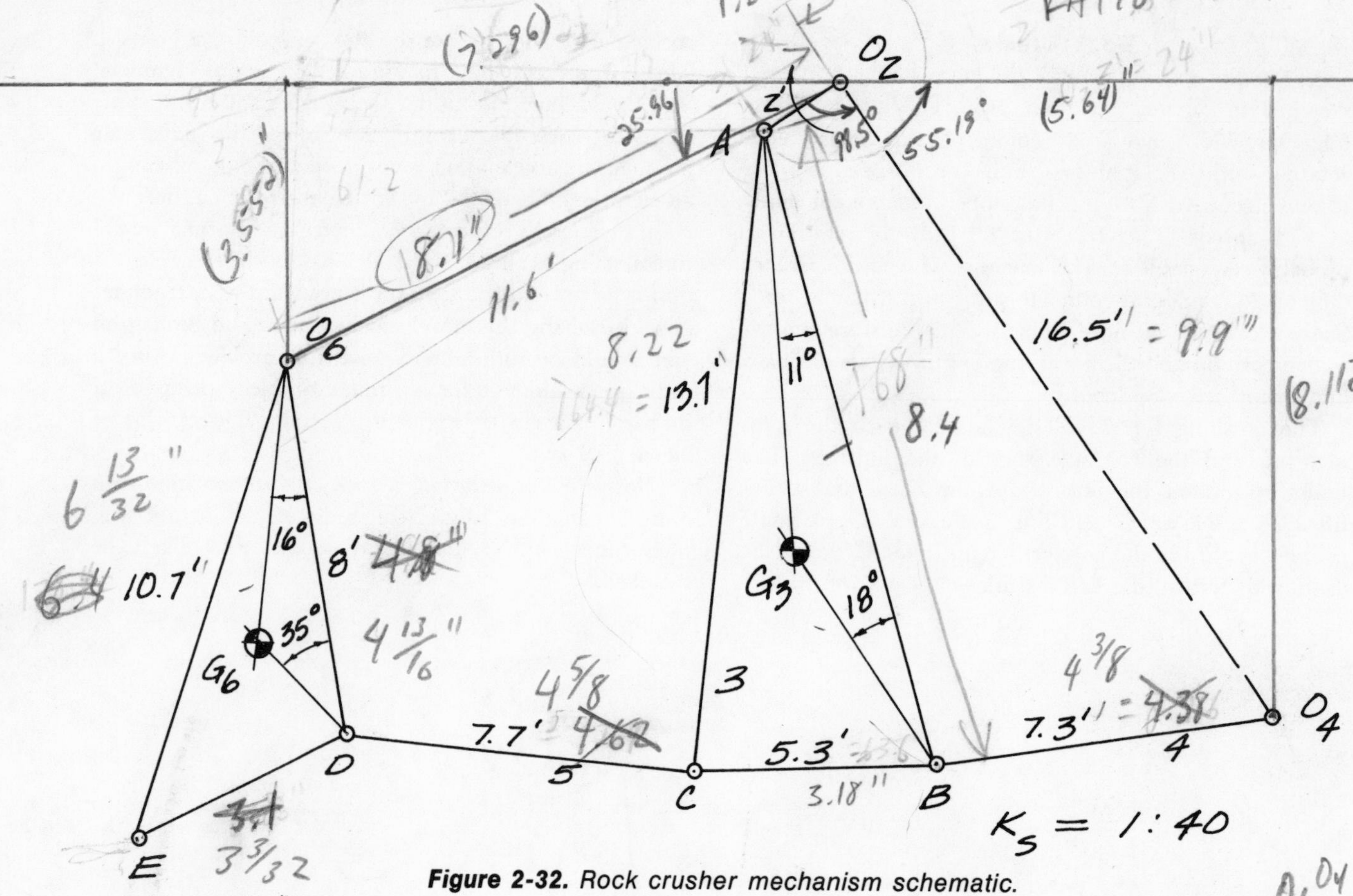

Figure 2-32. Rock crusher mechanism schematic.

Graphical Solution

Velocities

$$\overset{VO}{V_B} = \overset{VV}{V_A} \mathbin{+\!\!\!\!\rightarrow} \overset{VO}{V_{B/A}}$$

$$\overset{VO}{V_B} = \overset{VV}{V_C} \mathbin{+\!\!\!\!\rightarrow} \overset{VO}{V_{B/C}}$$

Accelerations

$$\overset{VV}{A_B}{}^{n} \mathbin{+\!\!\!\!\rightarrow} \overset{VO}{A_B}{}^{t} = \overset{VV}{A_A} \mathbin{+\!\!\!\!\rightarrow} \overset{VV}{A^{n}_{B/A}} \mathbin{+\!\!\!\!\rightarrow} \overset{VO}{A^{t}_{B/A}}$$

$$\frac{(117)^2}{7.3} \qquad 7{,}200 \qquad \frac{(46)^2}{14}$$

1,875.2 ft/sec² 7,200 ft/sec² 151.14 ft/sec²

$$A_C = A_A \mathbin{+\!\!\!\!\rightarrow} A^{n}_{C/A} \mathbin{+\!\!\!\!\rightarrow} A^{t}_{C/A} = A_B \mathbin{+\!\!\!\!\rightarrow} A^{n}_{B/C} \mathbin{+\!\!\!\!\rightarrow} A^{t}_{B/C}$$

$$\frac{(45)^2}{13.7} \qquad \frac{(17)^2}{5.3}$$

147.4 54.53

$$\overset{VV}{A_D}{}^{n} \mathbin{+\!\!\!\!\rightarrow} \overset{VO}{A_D}{}^{t} = \overset{VV}{A_C} \mathbin{+\!\!\!\!\rightarrow} \overset{VV}{A^{n}_{D/C}} \mathbin{+\!\!\!\!\rightarrow} \overset{VO}{A^{t}_{D/C}}$$

$$\frac{(27)^2}{8} \qquad \frac{(103)^2}{7.7}$$

91.12 ft/sec² 1,377.8 ft/sec²

$$\overset{VV}{A_E}{}^{n} \mathbin{+\!\!\!\!\rightarrow} \overset{VO}{A_E}{}^{t} = \overset{VV}{A_D} \mathbin{+\!\!\!\!\rightarrow} \overset{VV}{A^{n}_{E/D}} \mathbin{+\!\!\!\!\rightarrow} \overset{VO}{A^{t}_{E/D}}$$

$$\frac{(37)^2}{107} \qquad \frac{(18)^2}{5}$$

128 ft/sec² 64.8 ft/sec²

Link 6

$$s = \left(\frac{a + b + c}{2}\right) = \left(\frac{6 + 9 + 11.5}{2}\right) = 13.25 \text{ ft}$$

$$\text{Area} = \sqrt{s\,(s - a)\,(s - b)\,(s - c)}$$

$$= \sqrt{13.25\ (7.25)\ (4.25)\ (2.25)}$$

$$= 30.31 \text{ ft}^2$$

$$\text{Volume} = (30.31)\left(\frac{1}{12}\right)$$

$$= 2.56 \text{ ft}^3$$

$$\text{Weight} = (2.56)\ (490)$$

$$= 1{,}238 \text{ lb}$$

(Text continued on page 48)

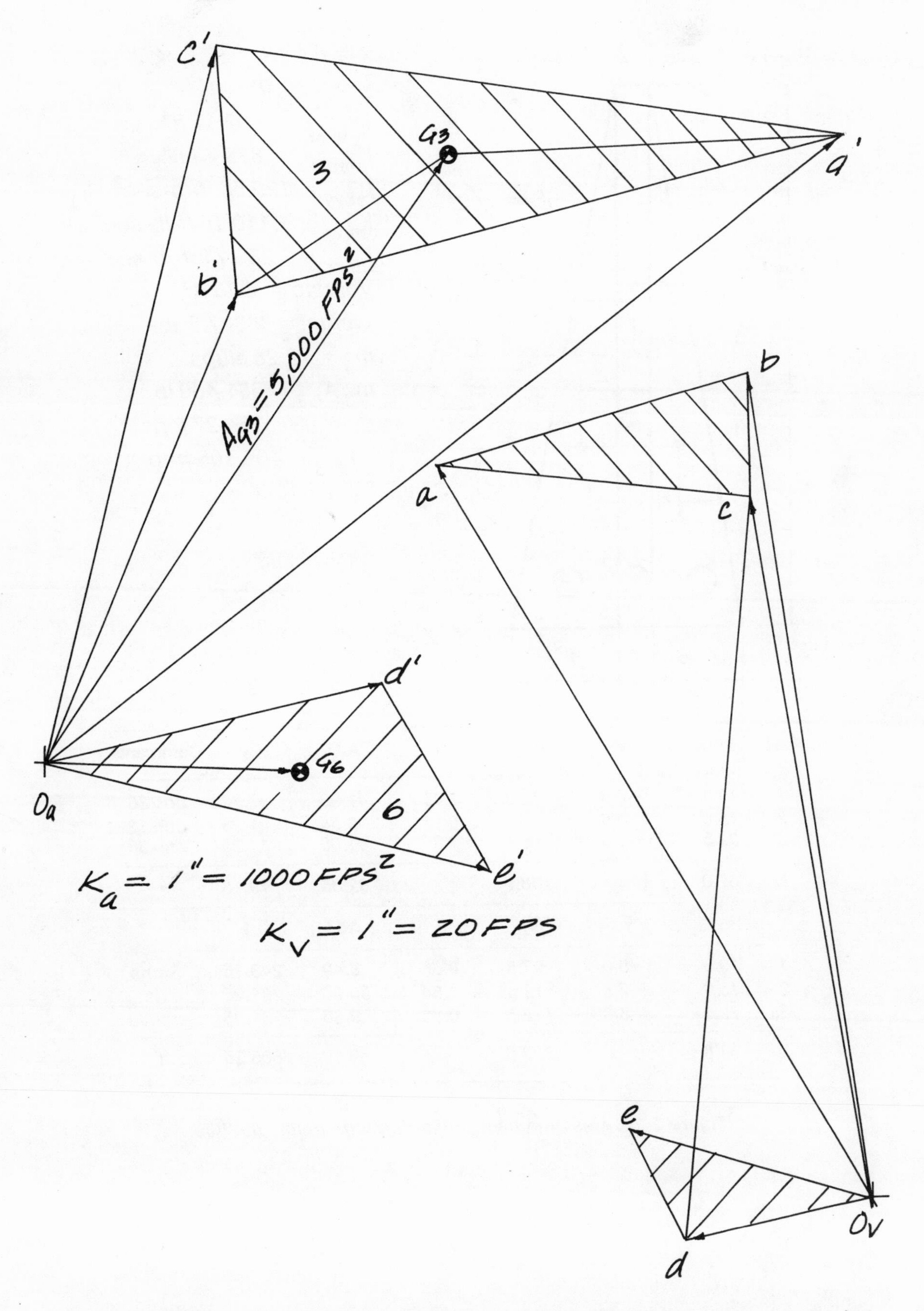

Figure 2-33. *Velocity and acceleration polygons for one position of a six-bar mechanism.*

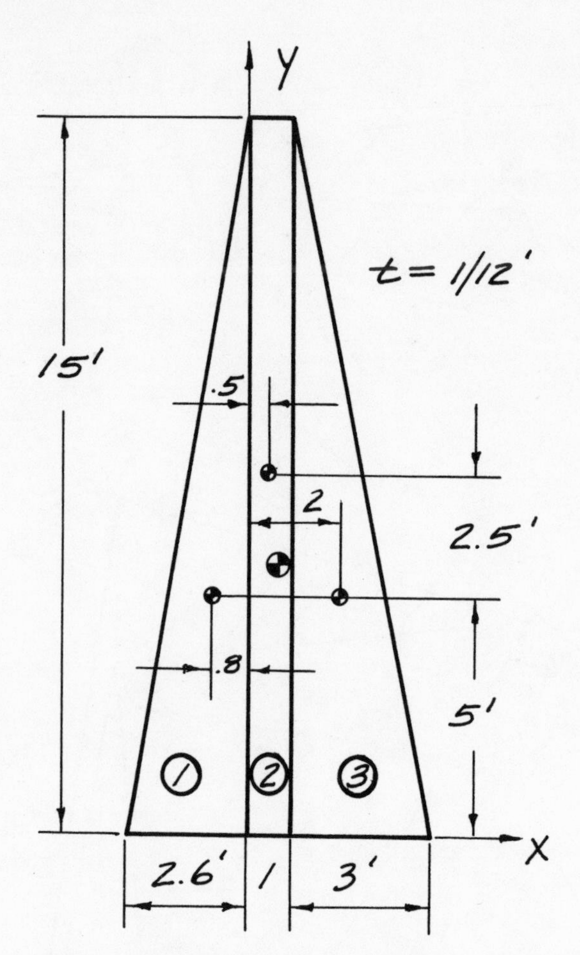

$\bar{x} = 0.65\ ft$

$\bar{y} = 5.66\ ft$

$I_{area,\ yy} = 102.17\ ft^4$

$I_{area,\ xx} = 875.44\ ft^4$

$I_{m,yy} = 129.56\ ft\text{-}lb\text{-}sec^2$

$I_{m,xx} = 1{,}110.16\ ft\text{-}lb\text{-}sec^2$

$I_{m,zz} = 1{,}239.71\ ft\text{-}lb\text{-}sec^2$

$Volume = 4.75\ ft^3$

$Weight = 2{,}327.5\ lb$

$m_3 = 72.28\ slugs$

$m_3 A_{G3} = 361{,}400\ lb$

$= 164.27\ tons$

$-I_3\alpha_3 = 409{,}105\ ft\text{-}lb$

$h_3 = 1.13\ ft$

Part	Area	$\bar{x}_i$	$A\bar{x}_i$	d_i	Ad_i^2	$\bar{I}$	Comments
1	19.5	−0.8	− 15.6	1.45	41.00	7.32	$bh^3/36$
2	15.0	0.5	7.5	0.15	0.34	1.25	$bh^3/12$
3	22.5	2.0	45.0	1.35	41.00	11.25	$bh^3/36$
Σ	57.0		36.9		82.35	19.18	I_{yy}
		$\bar{y}_i$	$A\bar{y}_i$	d_i	Ad_i^2	$\bar{I}$	
1	19.5	5.0	97.5	0.66	8.49	243.75	Same
2	15.0	7.5	112.5	1.84	50.90	281.25	as
3	22.5	5.0	112.5	0.66	9.80	281.25	above
Σ	57.0		322.5		69.19	806.25	I_{xx}

Figure 2-34. *Mass moment of inertia by graphical method.*

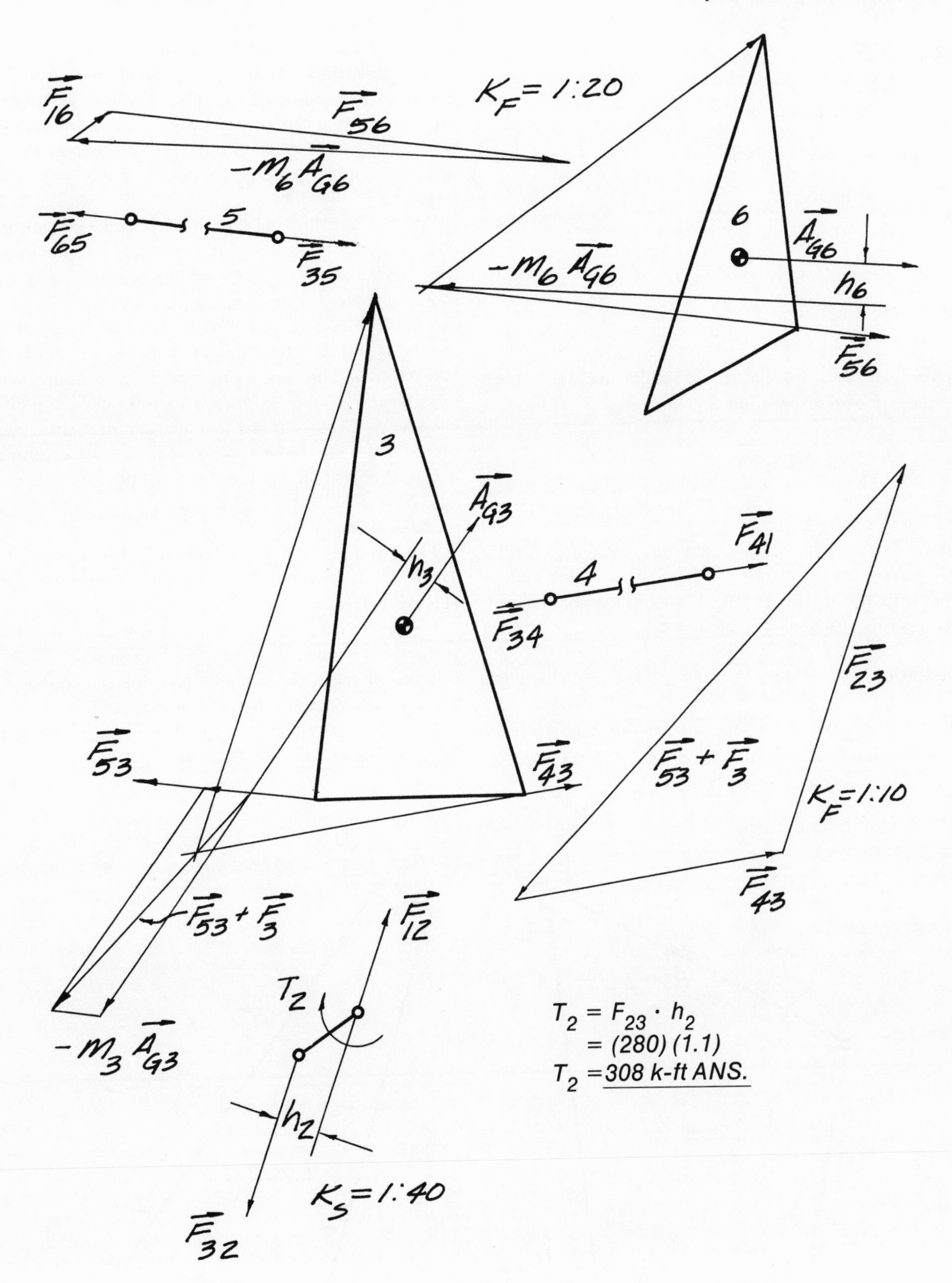

$$T_2 = F_{23} \cdot h_2$$
$$= (280)\,(1.1)$$
$$T_2 = \underline{308 \text{ k-ft ANS.}}$$

Figure 2-35. *Dynamic force analysis.*

$$\text{Mass} = \frac{W}{32.2}$$

$$= 38.44 \text{ lb-sec}^2/\text{ft}$$

$$-m_6 A_{G6} = (38.44)\,(1{,}800)$$

$$= 69{,}185 \text{ lb}$$

$$= 31.5 \text{ ton}$$

$$-\bar{I}_6 \alpha_6 \cong (250)\,(300)$$

$$= 75{,}000 \text{ ft-lb}$$

Where, $\bar{I}_6$ has been calculated in a similar manner as for the other triangular shape link 3. Therefore,

$$h_6 = \frac{75{,}000}{69{,}185} = 1.08 \text{ ft}$$

as shown in Figure 2-34.

(Note: The weight of the links has been neglected, since it accounts for a relatively small error when compared with the inertia effects. It should be considered when bearing reactions are calculated.)

Kinematics

The first step is to conduct a kinematic analysis. If you want to sudy just one position of the mechanism, the easiest way would be to draw graphically the velocity and acceleration polygons as shown in Figure 2-33. Knowing the angular velocities and accelerations for each link and the positions for each link, the accelerations of the centers of gravity can be found, and the dynamic analysis can be conducted by any chosen method.

For a complete cycle analysis, a program must be written in order to expedite the kinematic analysis of this mechanism. Any five- or six-bar mechanism, as shown, is much more complicated than a four-bar linkage. For this reason, a method of attacking such a task is shown. It should be made clear that this may not be the best solution. The Raven's method of determining velocity and acceleration by means of independent position equations[4] is chosen for this mechanism, although it turned out to be quite tedious as far as the determination of the position equations is concerned for this problem. Nevertheless this seems to be the best approach for this course level.

The geometry of the mechanism will be described by means of position vector in complex numbers form. (Figure 2-32)

Figure 2-36 shows the mechanism in terms of position vectors. To obtain the velocity and acceleration matrices in terms of the variable angles and link dimensions, two loops are considered (shown in Figure 2-36).

Loop 1 is a well known four-bar linkage representation. To simplify the calculations, all data for this portion

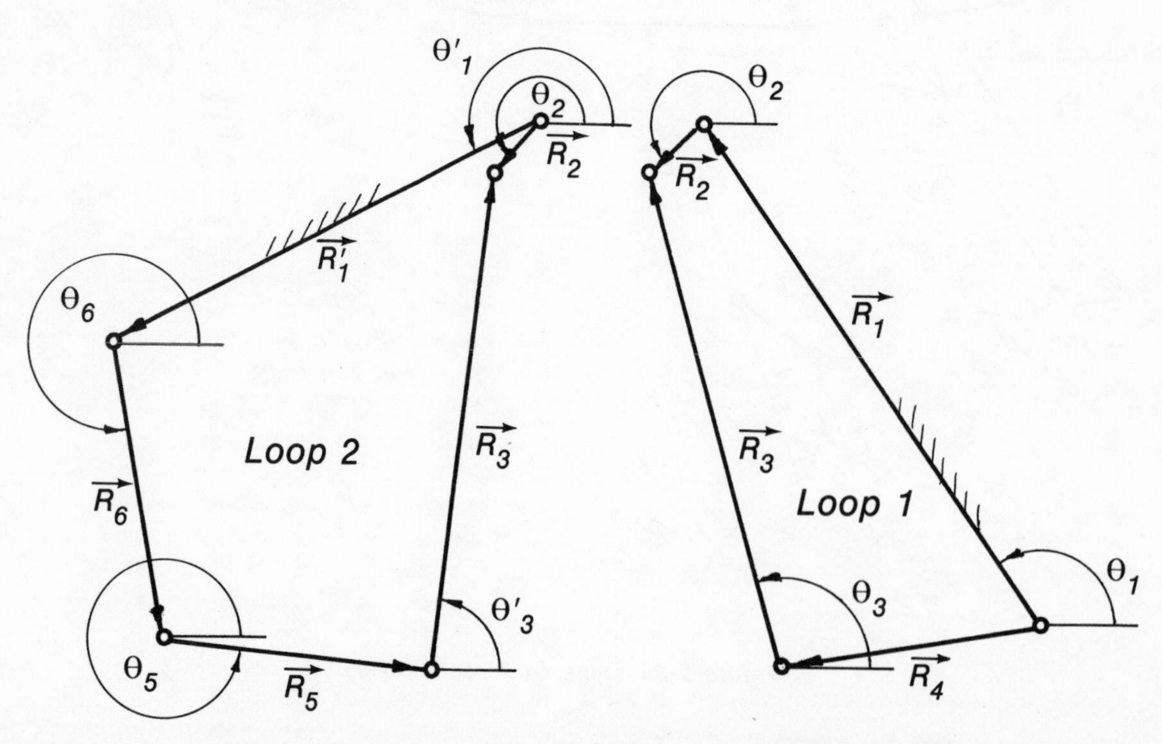

Figure 2-36. Schematic loop representation of a six-bar linkage.

could be obtained from the program included in Appendix 1.

Loop 2 is a five-bar mechanism linkage representation, and it needs to be solved completely.

The equations for Loop 1 and Loop 2 are:

Loop 1—

$$\vec{R}_1 + \vec{R}_2 = \vec{R}_4 + \vec{R}_3$$

Loop 2—

$$\vec{R}_2 = \vec{R}_1' + \vec{R}_6 + \vec{R}_5 + \vec{R}_3'$$

The two loop equations will be rewritten in the following form:

$$R_1e^{j\theta_1} + R_2e^{j\theta_2} = R_4e^{j\theta_4} + R_3e^{j\theta_3}$$

$$R_2e^{j\theta_2} = R_1'e^{j\theta_1'} + R_6e^{j\theta_6} + R_5e^{j\theta_5} + R_3'e^{j\theta_3'}$$

Next, these equations will be differentiated twice in order to obtain the required velocity and acceleration matrices.

Velocity Matrix

$$\begin{bmatrix} R_3\cos\theta_3 & R_4\cos\theta_4 & 0 & 0 \\ R_3\sin\theta_3 & R_4\sin\theta_4 & 0 & 0 \\ R_3'\cos\theta_3' & 0 & R_5\cos\theta_5 & R_6\cos\theta_6 \\ R_3'\sin\theta_3' & 0 & R_5\sin\theta_5 & R_6\sin\theta_6 \end{bmatrix} \begin{bmatrix} \omega_3 \\ \omega_4 \\ \omega_5 \\ \omega_6 \end{bmatrix} = \begin{bmatrix} R_2\omega_2\cos\theta_2 \\ R_2\omega_2\sin\theta_2 \\ R_2\omega_2\cos\theta_2 \\ R_2\omega_2\sin\theta_2 \end{bmatrix}$$

Acceleration Matrix

$$\begin{bmatrix} -R_3\sin\theta_3 & -R_4\sin\theta_4 & 0 & 0 \\ R_3\cos\theta_3 & R_4\cos\theta_4 & 0 & 0 \\ -R_3'\sin\theta_3' & 0 & -R_5\sin\theta_5 & -R_6\sin\theta_6 \\ R_3'\cos\theta_3' & 0 & R_5\cos\theta_5 & R_6\cos\theta_6 \end{bmatrix} \begin{bmatrix} \alpha_3 \\ \alpha_4 \\ \alpha_5 \\ \alpha_6 \end{bmatrix} = \begin{bmatrix} A \\ B \\ C \\ D \end{bmatrix}$$

where:

$$A = (R_2\omega_2^2\cos\theta_2) + (R_3\omega_3^2\cos\theta_3) + (R_4\omega_4^2\cos\theta_4)$$

$$B = (R_2\omega_2^2\sin\theta_2) + (R_3\omega_3^2\sin\theta_3) + (R_4\omega_4^2\sin\theta_4)$$

$$C = (R_2\omega_2^2\cos\theta_2) + (R_3'\omega_3^2\cos\theta_3') + (R_5\omega_5^2\cos\theta_5) + (R_6\omega_6^2\cos\theta_6)$$

$$D = (R_2\omega_2^2\sin\theta_2) + (R_3'\omega_3^2\sin\theta_3') + (R_5\omega_5^2\sin\theta_5) + (R_6\omega_6^2\sin\theta_6)$$

The accelerations of the center of gravity could be found easily after the angular velocities and accelerations for each link are known. For example,

$$Ag_6 = R_{G6}\,[-\alpha_6\sin\theta_{RG6} - \omega_6^2\cos\theta_{RG6}]\vec{i} + R_{G6}\,[\alpha_6\cos\theta_{RG6} - \omega_6^2\sin\theta_{RG6}]\vec{j}$$

Substituting given data yields the following matrices and results:

Velocity Coefficient Matrix

$$\left[\begin{array}{cccc|c} -3.90 & -7.23 & 0 & 0 & 100.64 \\ 13.44 & -1.02 & 0 & 0 & 65.36 \\ 1.38 & 0 & 7.64 & 1.44 & 100.64 \\ 13.63 & 0 & -0.96 & -7.87 & 65.36 \end{array}\right]$$

or

$$\omega_3 = 3.66\ \frac{\text{rad}}{\text{sec}} \qquad \omega_4 = -15.89\ \frac{\text{rad}}{\text{sec}}$$

$$\omega_5 = 13.18\ \frac{\text{rad}}{\text{sec}} \qquad \omega_6 = -3.57\ \frac{\text{rad}}{\text{sec}}$$

Acceleration Coefficient Matrix

$$\left[\begin{array}{cccc|c} -13.44 & 1.02 & 0 & 0 & 4{,}160 \\ -\ 3.90 & -7.23 & 0 & 0 & 3{,}845 \\ -13.63 & 0 & 0.96 & 7.87 & 7{,}402 \\ 1.38 & 0 & 7.64 & 1.48 & 3{,}836 \end{array}\right]$$

or

$$\alpha_3 = -336.08;\ \frac{\text{rad}}{\text{sec}^2} \qquad \alpha_4 = -350.52;\ \frac{\text{rad}}{\text{sec}^2}$$

$$\alpha_5 = 506.88;\ \frac{\text{rad}}{\text{sec}^2} \qquad \alpha_6 = 296.69\ \frac{\text{rad}}{\text{sec}^2}$$

In order to prepare the data and equations for programming, the next step is to express all angles in terms of θ_2 and the given link dimensions. Also, a force matrix needs to be set up in order to solve the forces and torques needed. The force matrix will have a similar form to that in the previously worked problem.

It would be a challenge to write such a program for a pocket calculator, but it appears that this is beyond the capability of the currently marketed programmable hand calculators due to insufficient memory allocations.

Computer Aided Design Programs

Engineers today are in a better position to solve general dynamics problems because of the appearance of digital computers. Many machines contain "floating links" with variable inertia forces, which result in nonlinear differential equations of motion which must be solved. Until recently, most of the dynamic force analyses were conducted as shown in the graphical methods included in this text. These methods of solution are useful in understanding the principles involved, but would be of little, if any, value today.

Again, it is beyond the scope of this text to discuss in detail the various general purpose computer programs for the solution of linkages.

Programs such as IMP, DRAM, MEDUSA, VECNET, ADAMS and recently, DISCOS do the following:

1. Generate the equations of motions of machines (differential eq. with $m\ddot{x}$, $I\ddot{\theta}$ etc. terms)
2. Solve the equations (integrate them numerically)
3. Determine the reactions at the joints, torques, bearing forces, etc.
4. When combined with plotters, display graphical results

It is important to realize that the solutions done analytically differ significantly from the known graphical methods. In fact, some recommend that graphically oriented thinking be abandoned when using them. In order to use these programs, you do not necessarily need to be an expert in dynamics, but it certainly would help to understand the principles. IMP, DRAM, DISCOS and other programs use the Lagrange's equations in the analysis of mechanisms. Methods for solving the differential equations of motion (e.g., the Runge-Kutta) were developed and are available as subroutines in most computer programs today.

A brief description of three of the programs follows. This is the state-of-the-art for any serious designer of machinery. Computer-aided designs are powerful tools in the hands of a designer. This is also true for anyone who wants to be competitive in industry. It does little good to study out-of-date dynamics books, except for learning the basics.

Integrated Mechanism Program

IMP, Integrated Mechanism Program, is a computer-aided design analysis system for mechanisms and linkages.[5] IMP was developed by P. N. Sheth and Professor J. J. Uicker, Jr. of the University of Wisconsin at Madison. By 1972, it had been released for general distribution to 35 U.S. manufacturing industries, 29 universities, 13 computer hardware and software companies, and 10 organizations outside the United States. It took approximately 10 years to develop this program. The National Science Foundation, Ford Motor Company, and others contributed money for completion of the task.

IMP uses FORTRAN IV. The program is quite long, approximately 200 pages of printout, and the output can be either printed or a graphic display.

IMP is intended to aid in the analysis of two- or three-dimensional rigid link mechanisms having single or multiple degrees of freedom including revolute, prismatic, screw, spur gear, cylindric, universal, spheric, and planar joints in any closed loop combination.

The program is capable of kinematic, static, and dynamic analysis including vibration analysis. Again, it uses Lagrange's equation of motions and other means of network programming to accomplish its task. An excellent user's guide is available which includes examples of application (industrial sewing machine mechanism, a two-degree of freedom control linkage—7 links, 6-revolute pairs, and 1 sliding pair). A truck suspension mechanism and a landing gear mechanism of a jet aircraft would be good examples of application. IMP will calculate the desired positions, velocities, accelerations, the static and dynamic forces, and the natural frequencies of the mechanism for any input angle and print or display results.[5]

Dynamic Response of Articulated Machinery

DRAM, Dynamic Response of Articulated Machinery, was developed under the direction of Professor M. A. Chace of the University of Michigan at Ann Arbor in 1973.[5]

DRAM is particularly useful to mechanical systems with large scale displacements such as the shovel mechanism when it is lifting a load. There are some limitations on the use of this program. The applications are limited to two-dimensional problems, lower-pair joints, no-impact loads, closed loops, etc. DRAM uses FORTRAN IV language, and can produce either printed or graphic display output.

You can use the program without in-depth understanding of the details since an excellent user's guide is available. The theory of the program is quite involved—use of Lagrange equations, network theory, involved programming, etc. A graduate course in dynamics and mechanisms would be a necessity in order to master it. Nevertheless, this program would handle the problem presented here with ease.

To make sure that you realize the power of such a program, consider what it really does. In a matter of seconds the program will calculate and display or write all the forces, pin reactions, etc. for any position θ_2 as requested. Additionally, animation of the motion is possible by the use of available equipment. Therefore, the traditional method of kinematic-dynamic analysis by drawing velocity-acceleration-force polygons has only educational value.[5]

Dynamic Interaction Simulation of Controls and Structure

DISCOS, Computer Program System for Dynamic Simulation and Stability Analysis of Passive and Actively Controlled Spacecraft, was developed by the Dynamics and Loads Section, Martin Marrieta Corporation, Denver Division while under contract to NASA in 1975. The dynamic system is modeled as an assembly of rigid and/or flexible bodies. The computer program could be used to determine total dynamic system performance.

DISCOS is perhaps the most general program known. Justification of its use for the simple mechanisms problems given in this text would depend on availability and the cost involved. Most likely, you would use the simplest program depending on the number of links involved and the type of analysis needed. For example, there is no need to use a program which handles a three-dimensional case if only a two-dimensional analysis is required. Also, it would cost more. If the other is not available, however, you may wish to do so.

In the near future, mini-computers will most likely be in a position to handle problems of kinematics-dynamics analysis.

Review Problems

Problem 1

Find the force F required to balance the torque T_2 of 150 in.-lb. of the mechanism shown in the figure. Choose appropriate K_F scale for reasonable accuracy. Position free-body diagrams and force polygon in such a way that the drawings will not interfere and position them in a logical sequence of solution. Assume a K_s scale of 1:10.

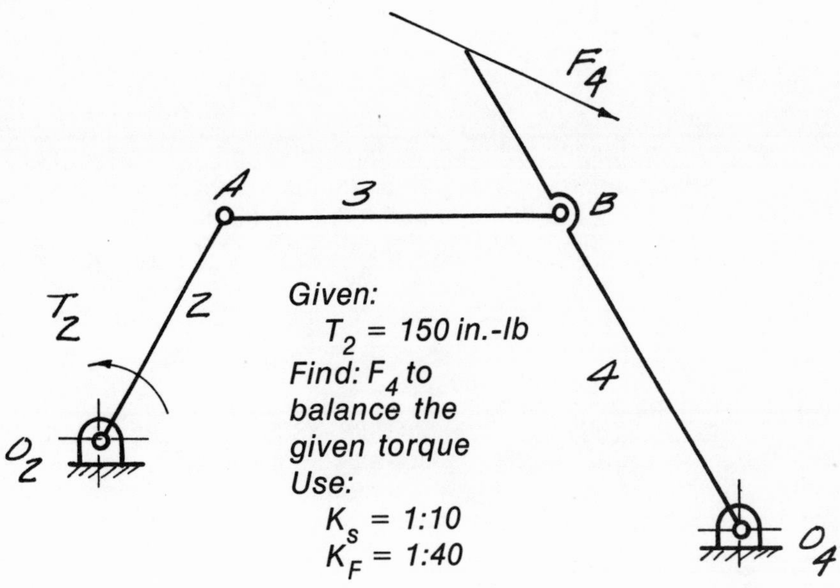

Problem 2

For the mechanism shown in the figure, determine the cylinder force required to balance a force on the link 6 equal to $F_6 = 1{,}000$ lb. Use a proper K_F for reasonable accuracy of drawing. K_S is 1:10.

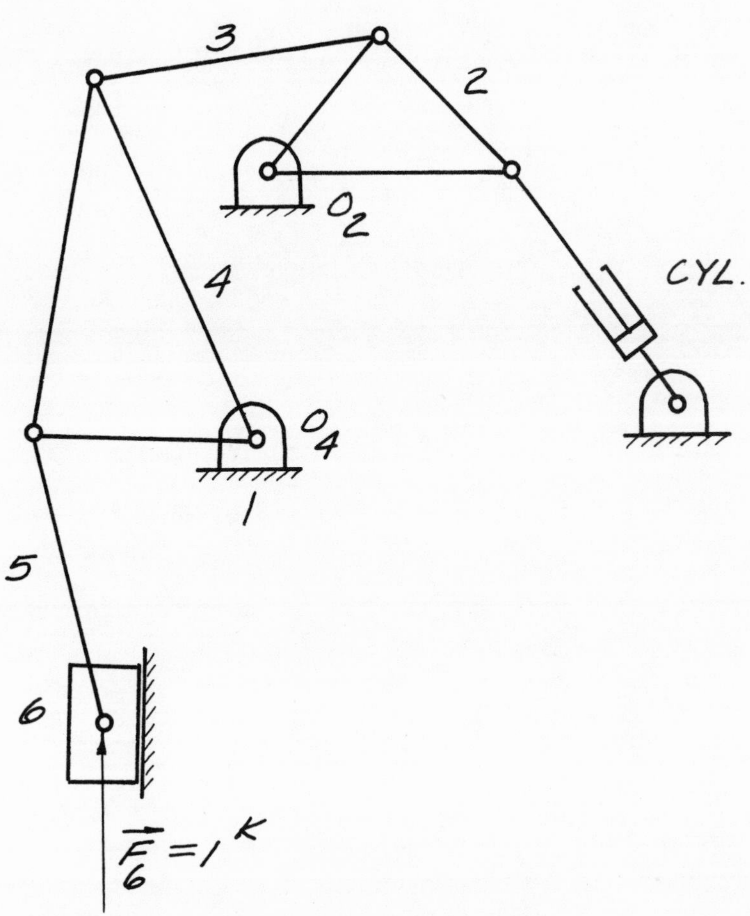

Problem 3

Find the torque T_2 and force F_{32} for the mechanism shown in the figure. Use scales: K_S 1″ = 10″; K_F: 1″ = 100 lb. Work graphically and analytically.

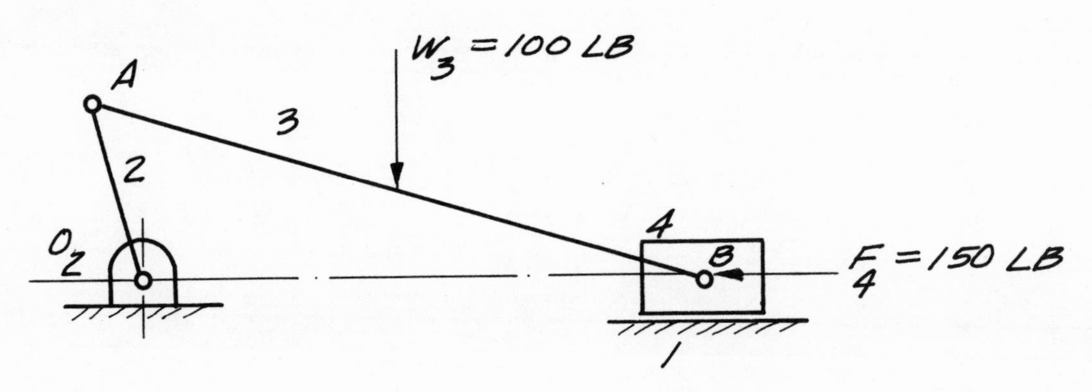

Problem 4

For the slider crank mechanism shown in the figure, do the following:

1. Graphically draw the force-analysis for $\theta_2 = 45°$, and 240°.
2. Determine analytically the equation for the torque T_2 in terms of the applied force F and dimensions shown.
3. Program the equation of part 2 on a pocket calculator, and make a graph of T_2 as a function of θ_2 for every 30° of increase of θ_2 assuming a given force $F = 1$ lb. Use other dimensions as read from the drawing by using a scale of $K_S = 1:10$.

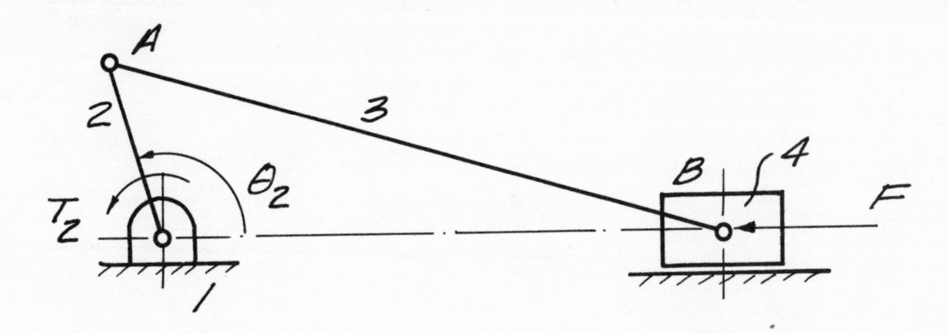

Problem 5

Two geared four-bar mechanisms (using 20° pressure angle spur gears) are shown in the figure. Both mechanisms use the same sizes for the driver and driven links. Determine the pin reactions F_{13} for each case assuming for each mechanism the same forces F_{23} and F_{43} as shown. Discuss the difference.

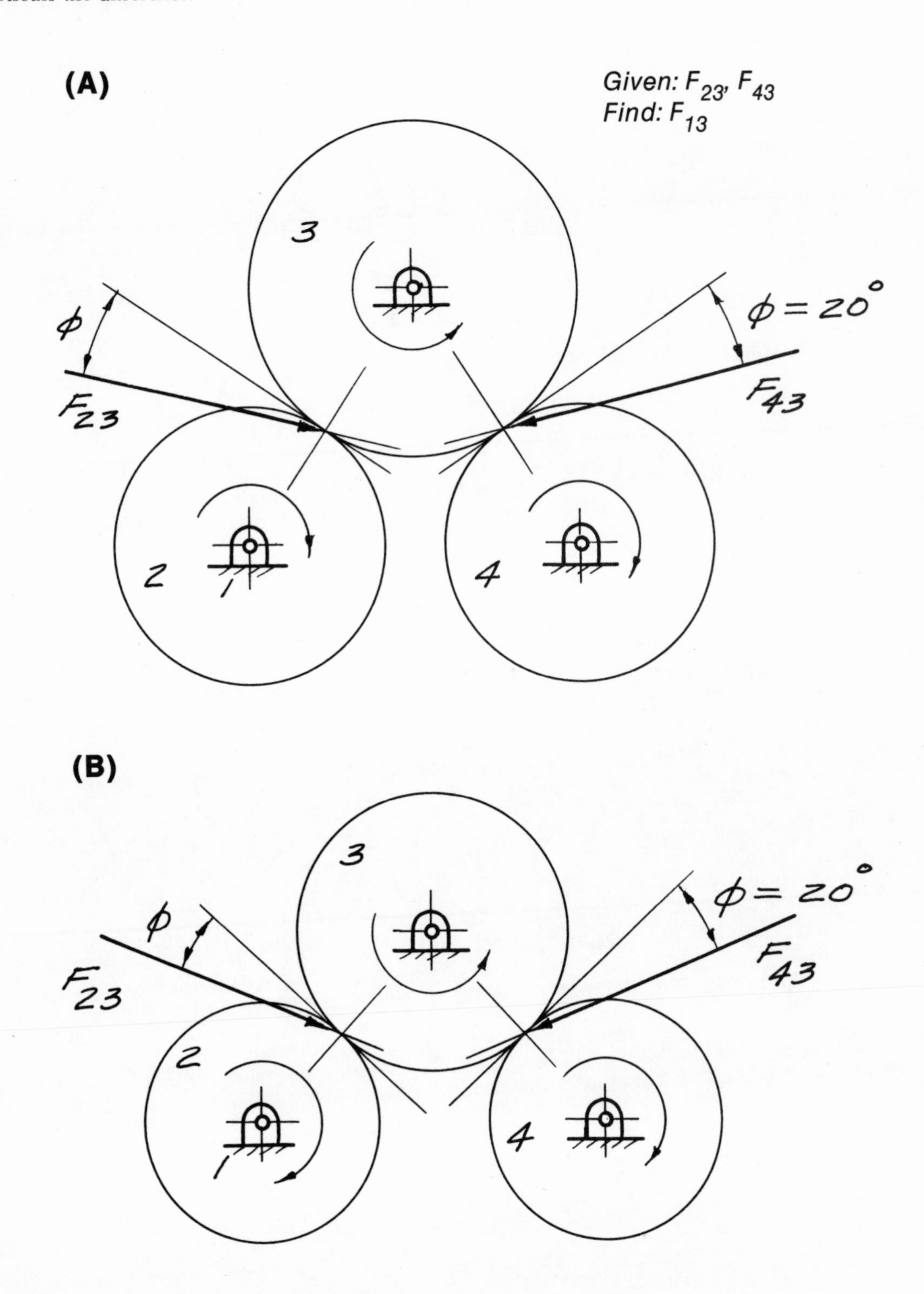

Problem 6

Take the data of the shovel mechanism and check the answers in Table 2-2 for any of the listed crank positions. Use graphical or analytical methods in solving this problem. If you have any difficulty, read the section again, and follow the given solution.

Problem 7

Using the data for the rock crusher as given in these notes, conduct a dynamic force analysis by including the weights of link 4 and link 5. Assume that the rods are made of steel and are 3 inches in diameter.

Problem 8

For any of the four static force analysis examples (Figures 2-14, 2-15, 2-16 and 2-17), do the following:

1. Assume dimensions needed
2. Assume any reasonable positive constant angular velocity for the driver ($\alpha_2 = 0$)
3. Assume masses of links (one or more)
4. Conduct a dynamic-force analysis using graphical techniques
5. Conduct a dynamic-force analysis using analytical methods
6. If possible, balance mechanism

Note: Use proper scales for best drawing accuracy, present a well-organized solution.

Problem 9

Given:

$\omega_2 =$ ______ rad/sec ccw, const.
$\theta_2 =$ ______ deg
$P_1 =$ ______ lb
$P_2 =$ ______ lb
$O_2A =$ ______ in.
$AB =$ ______ in.
$BO_4 =$ ______ in.
$O_2O_4 =$ ______ in.
$W_3 =$ ______ lb
$W_4 =$ ______ lb
$r_{G3} =$ ______ in.
$I_3 =$ ______ lb in.-sec^2
$I_4 =$ ______ lb in.-sec^2
$r_{G4} =$ ______ in.

Results::

$F^{I}_{23,\ m_3}$	$F^{II}_{23,\ m_4}$	$F^{III}_{23,\ P_1}$	$F^{IV}_{23,\ P_2}$	ΣF_{23}	T_2
lb	lb	lb	lb	lb	in.-lb

Find: (Graphically and Analytically)

$\omega_3 =$ ______ rad/sec
$\omega_4 =$ ______ rad/sec
$A_{G3\vec{i}} =$ ______ in./sec^2
$A_{G4\vec{i}} =$ ______ in./sec^2
$h_3 =$ ______ in.
$\alpha_3 =$ ______ rad/sec^2
$\alpha_4 =$ ______ rad/sec^2
$A_{G3\vec{j}} =$ ______ in./sec^2; $\theta_{AG3} =$ ______°
$A_{G4\vec{j}} =$ ______ in./sec^2; $\theta_{AG4} =$ ______°
$h_4 =$ ______ in.

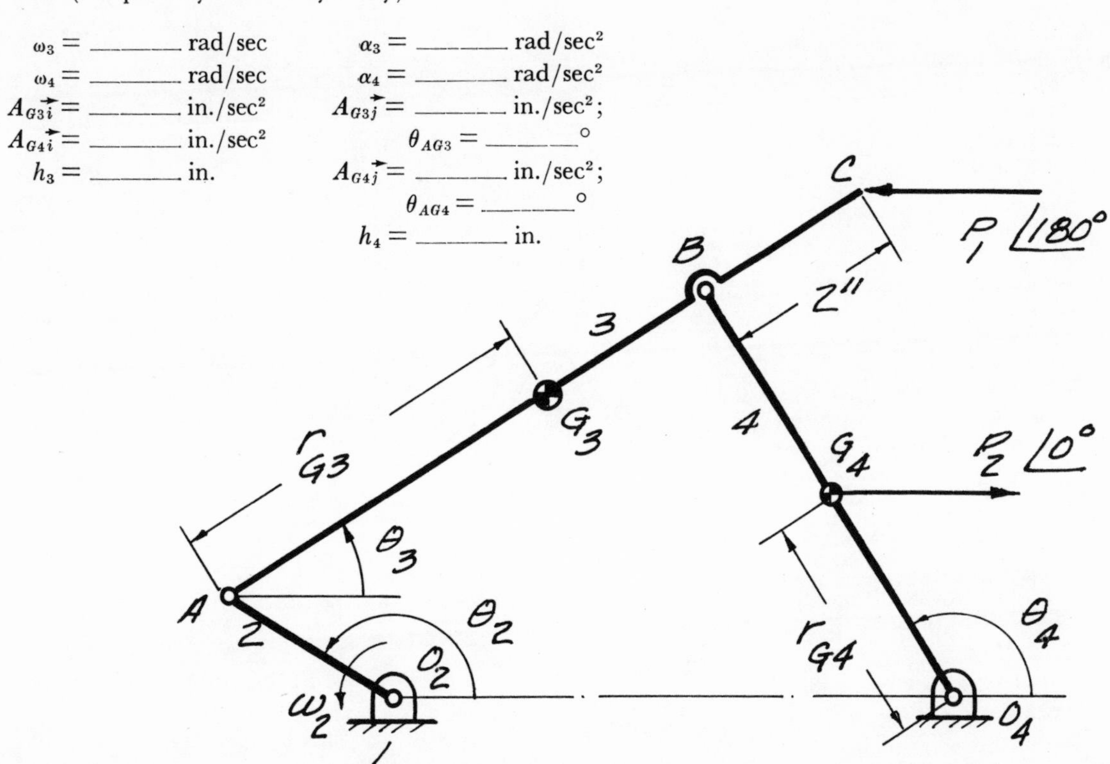

Problem 10

For the mechanism shown in the figure:

1. Draw the velocity and acceleration polygons
2. Conduct a dynamic force analysis finding the torque T_2 and shaking forces. Draw a free-body diagram for each link separately including weights of links.
3. Work the problem analytically. Show all work.

Data:

$m_2 = 2.5$ kg; $m_3 = 0$; $m_4 = 1.5$ kg (assume masses uniformly distributed)

$W_2 = 2.5\ (9.81)\ \text{kg}\cdot m/\text{sec}^2 = 24.5$ N (Newtons)

$W_3 = 0$; $W_4 = 14.7$ N; $T_4 = 100$ N cm; dimensions as shown.

$\omega_2 = 20$ rad/sec; ccw, constant

Choose appropriate scales for reasonable drawing accuracy.

$K_v: 1'' = 250$ cm/sec

$K_a: 1'' = 5{,}000$ cm/sec^2

$K_F: 1'' = 50$ N

or if you have a metric scale, use

K_v: 1 cm = 100 cm/sec, etc.

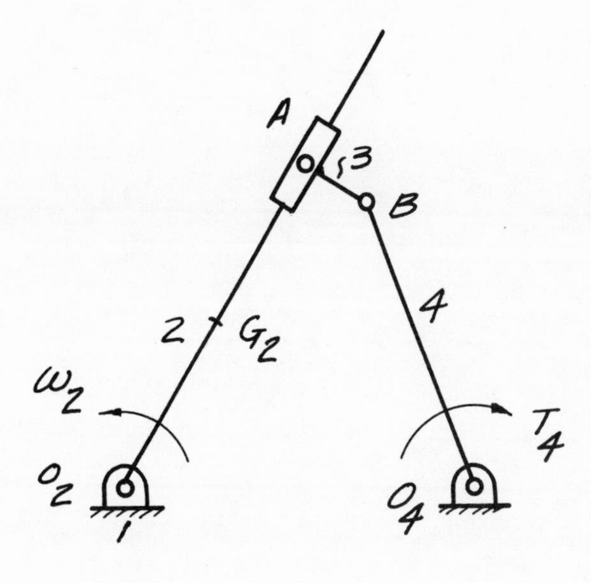

Problem 11

For the belt loop cutter shown, conduct a similar dynamic analysis to that of the shovel mechanism. For this purpose:

1. Study the motions of each link
2. Draw velocity and acceleration polygons for one arbitrary position of this mechanism
3. Draw static force-polygons
4. Draw dynamic force-polygon
5. Outline procedures for complete dynamic force analysis

Assume suitable external loads.

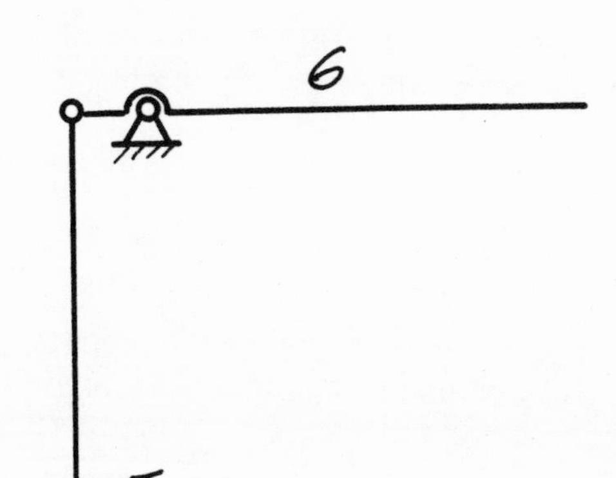

5

K_s ; 1″ = 2″
K_s = 1:20

4

1

3

2

Problem 12

Assuming suitable weight, load, conduct a dynamic analysis of the five-link bar called wobble knife mechanism shown in the figure. It is used in combines to cut the stalks of grain plants which are harvested. As the knife oscillates back and forth between the guides, it is allowed to move slightly to the front and rear; hence the name wobble knife. Given: $\omega_2 = 50 \ \frac{\text{rad}}{\text{sec}}$ cw, const., dimensions as shown. (Note: The newer combines use a gear drive instead of a four-bar linkage.)

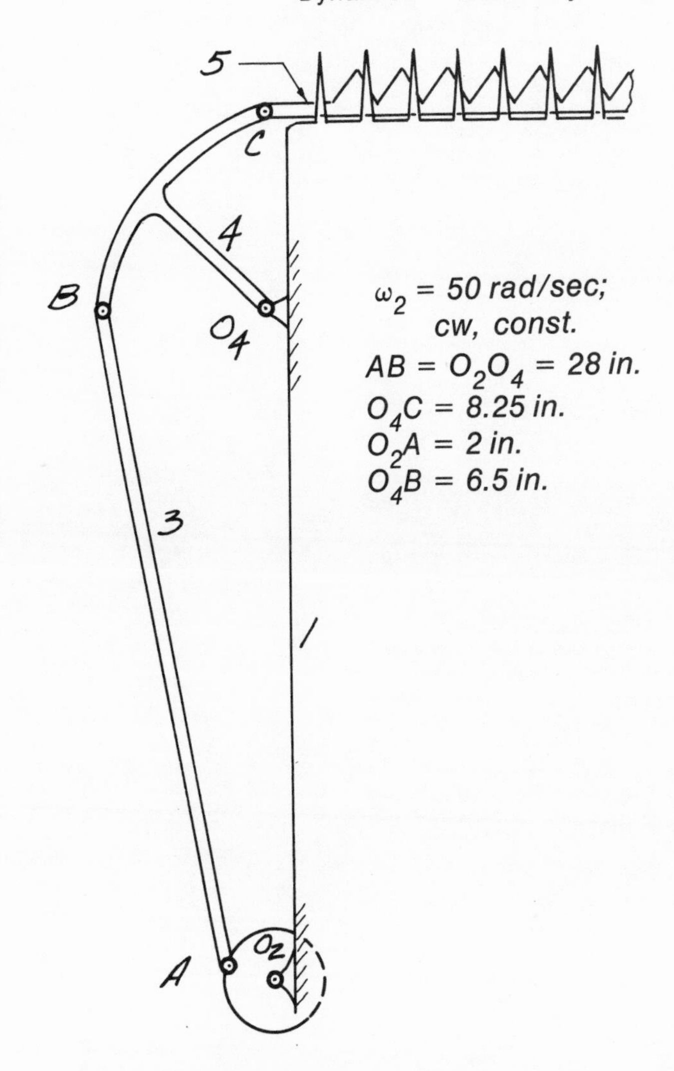

Problem 13

Given:

$W_5 = 1$ lb

$W_3 = 0.5$ lb

$\omega_2 = 1$ rad/sec, cw, const.

$K_S = 1:1$

Find: Will follower separate from link 2? Use equivalent mechanism. (Hint: Force F_{25} down)

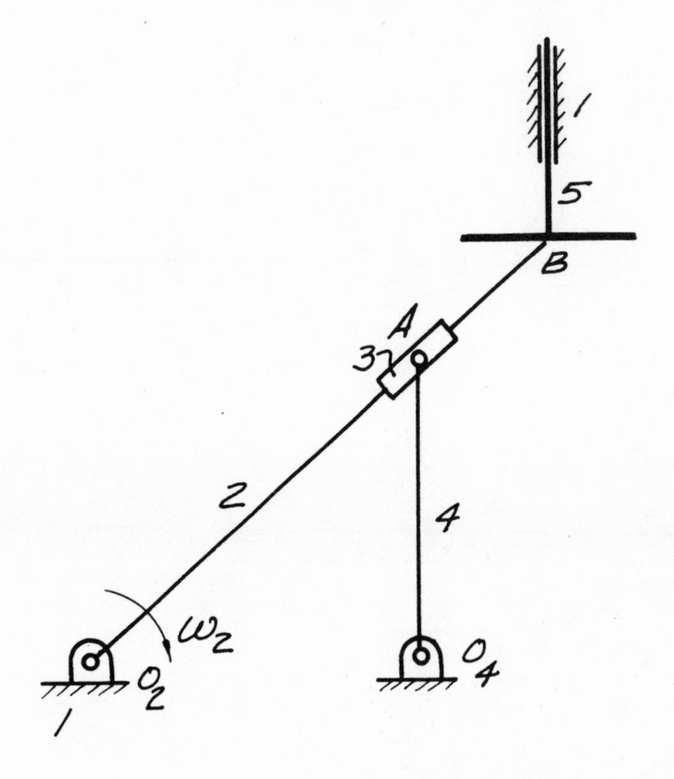

Problem 14

Mechanisms:

1. Inverted slider-crank, Type II
2. Offset slider-crank (See Figure 1-8)
3. Any five-link mechanism

For the mechanisms named, do the following:

1. Briefly describe different methods for determining forces and torques for the driver links
2. Develop formulas for linear and angular accelerations of different links by use of complex numbers (similar to those in the examples) and write a program to be executed on a pocket calculator
3. Check answers of part 2 by graphical methods for one position of the driver link
4. Write formulas and a program including the dynamic force analysis, check your answers by graphical procedures

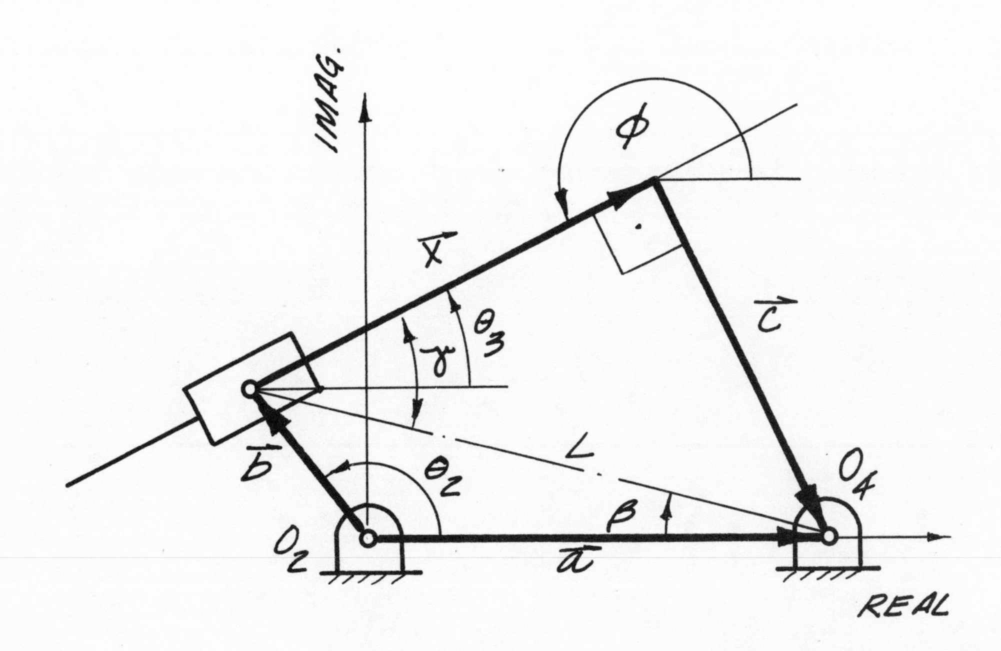

3
Balancing of Machinery

Static and Dynamic Balancing of Rotating Rigid Bodies

In the problems done before, the effect of the inertia forces (causing shaking forces) was obviously unwanted. The question arises of how to correct this problem. It is possible to balance the inertia forces by adding inertia forces which are equal in magnitude and acting in an opposite direction.

For example, the unbalance of crank 2 in Figure 3-1 could be corrected by adding weight, W_2, at the same distance as W_1, Figure 3-2.

It is difficult to make a perfectly aligned shaft, especially at a reasonable cost. Sometimes it is better to produce machine parts which might be "off center" and to correct the unbalance on a balancing machine. There are two types of balancing, static and dynamic. Other classifications would refer to the balancing of reciprocating (slider) or of rotating machinery (shafts). In this brief outline only rotating bodies are discussed.

To show the difference between static and dynamic balancing, consider a balanced rotor resting on two horizontal knife-edges as shown in Figure 3-2. You could mark the rotor with paint or scratch a mark on it, give it a light spin and then let it roll freely until it comes to rest. After repeating this, the mark will be found at a different place each time it stops. The rotor is considered to be *unbalanced* when the mark always stops at the same position.

If the rotor always stops at random positions, it is said to be *statically balanced.* Notice that both reactions in Figure 3-2 will be in the same direction, up, and in the same vertical plane.

Now consider two masses attached to another rotating shaft mounted on two bearings at the ends A and B, as shown in Figure 3-3. Again, even though the weights are equal and the system is statically balanced at any position, the system actually is unbalanced. When such a shaft rotates, there will be a couple acting on the system trying to turn it end-to-end. The centrifugal force of the weight $\vec{W}$ sets up a couple which reacts with the frame. The forces are statically balanced but the resultant couple ($mr\omega^2a$) "requires" a bearing couple. The frame (or bearings) must exert a couple on the rotor equal and opposite to the couple $mr\omega^2a$, Figure 3-3.

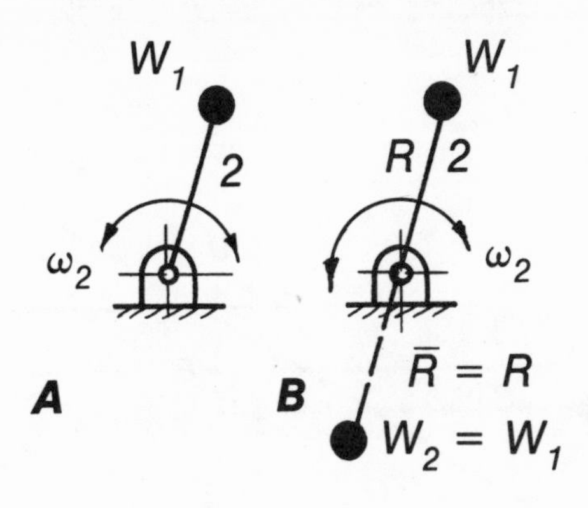

Figure 3-1. *Balancing of a rotating mass.*

These forces are variable loads acting on the frame and the phenomenon is called dynamic unbalance, which occurs only during motion of the shaft.

To make sure that there are no doubts about the difference in static and dynamic balancing, the analysis is repeated by using a different example. For this purpose consider a perfectly balanced automobile wheel mounted as shown in Figure 3-4 (step 1).

Next, attach a weight, $\vec{W}$, on top of one side of it as shown (step 2). Any minor disturbance will cause the wheel to roll until the weight is on the bottom.

If you take another weight, $\vec{W}$, (same magnitude as the first one) and apply it opposite to the first one, the wheel will not have the tendency to roll. This means that the wheel has been statically balanced.

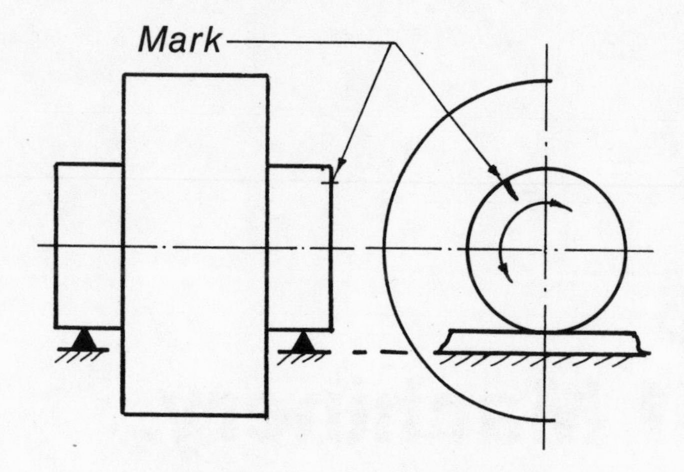

Figure 3-2. Marking of a rotor.

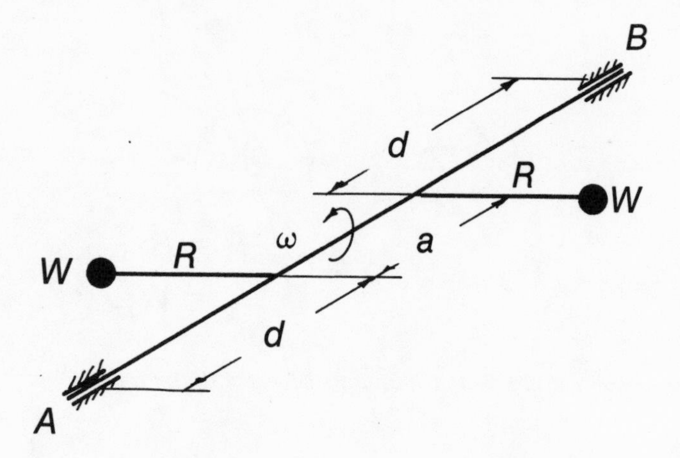

Figure 3-3. Two rotating masses.

What will happen if you remove the second weight and position it in the same plane but at an angle different than 180° from the first weight? It will become unbalanced again. If, on the other hand, you place the second weight in a different plane (parallel to the first plane) the system will become dynamically unbalanced, regardless of the angle relative to the first weight. If a rotor is balanced dynamically, it is also balanced statically, therefore, a complete balance means a dynamic balance.

As can be seen from the examples, when dealing with an unbalance of rotating rigid bodies, two distinct cases occur:

1. All rotating masses are in a single plane as shown in Figure 3-5
2. The rotating masses are in different planes as shown in Figure 3-6.

The determination of the required correction weights is easier in the first case. Only one equation must be solved either graphically or analytically, i.e., $\Sigma\vec{F} = 0$. In the second case, an additional equation must also be solved graphically or analytically, i.e., $\Sigma\vec{M} = 0$. These two equations are used in the examples solved.

You could also define static and dynamic balancing by the use of these two equations, i.e., as static balancing in the case of only the first equation, while dynamic balancing is the case which requires an additional moment equation for determining the correction weights.

Other characteristics of *static unbalance* are:

1. Static unbalance refers to an object at rest
2. It is due to the action of gravity
3. It can always be balanced by adding only one correction mass in any single plane

Dynamic unbalance is characterized by:

1. The action of the inertia forces taking place only when a body is rotating
2. A requirement that two planes must be used for correction (i.e., it is not possible to balance a moment by the use of one force only)

There are many techniques of balancing machinery, quite often involving very sophisticated equipment. But, this book is concerned with the basic theory of balancing rather than technology, and therefore, only the simplest problems will be covered.

There are two basic techniques for analysis of unbalance—graphical and analytical. The two known equations of equilibrium for a rigid body state:

$$\Sigma\vec{F} = 0$$

$$\Sigma\vec{M} = 0$$

or

$$\Sigma W\vec{R} = 0$$

$$\Sigma W\vec{R}d = 0$$

Note that for constant ω and g, the centrifugal force $(W/g)r\omega^2$ is proportional to the "WR" term. The "WR" term will be used as a measure of correction.

When written in a scalar form, the equations become:

$$\Sigma W_i R_i \cos\theta_i = 0$$

$$\Sigma W_i R_i \sin\theta_i = 0$$

and

$$\Sigma W_i R_i d \cos\theta_i = 0$$

$$\Sigma W_i R_i d \sin\theta_i = 0$$

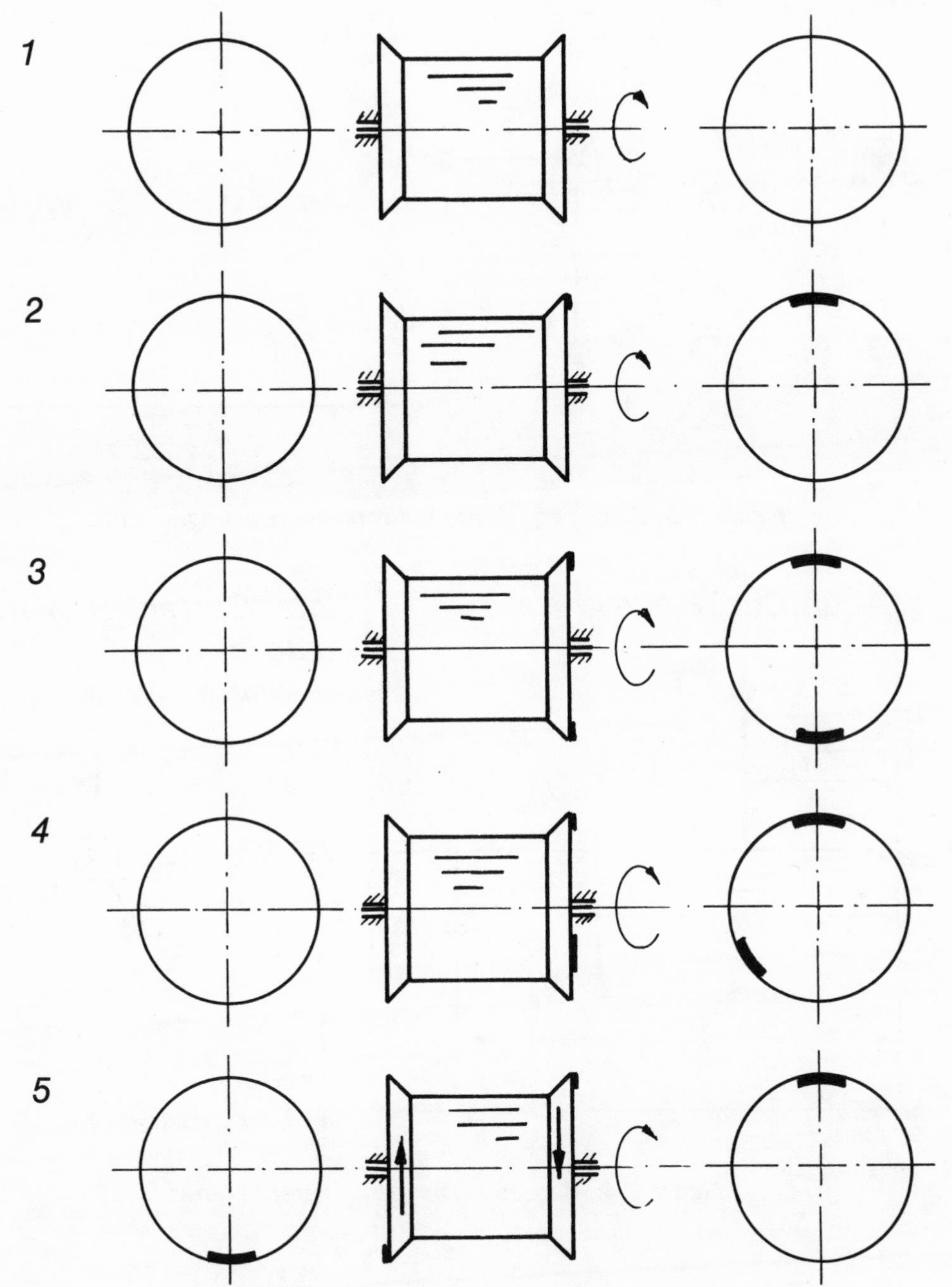

Step 1 — Statically and Dynamically Balanced
2 — Statically Unbalanced
3 — Statically Balanced
4 — Statically Unbalanced
5 — Dynamically Unbalanced

Figure 3-4. *Automobile wheel balancing.*

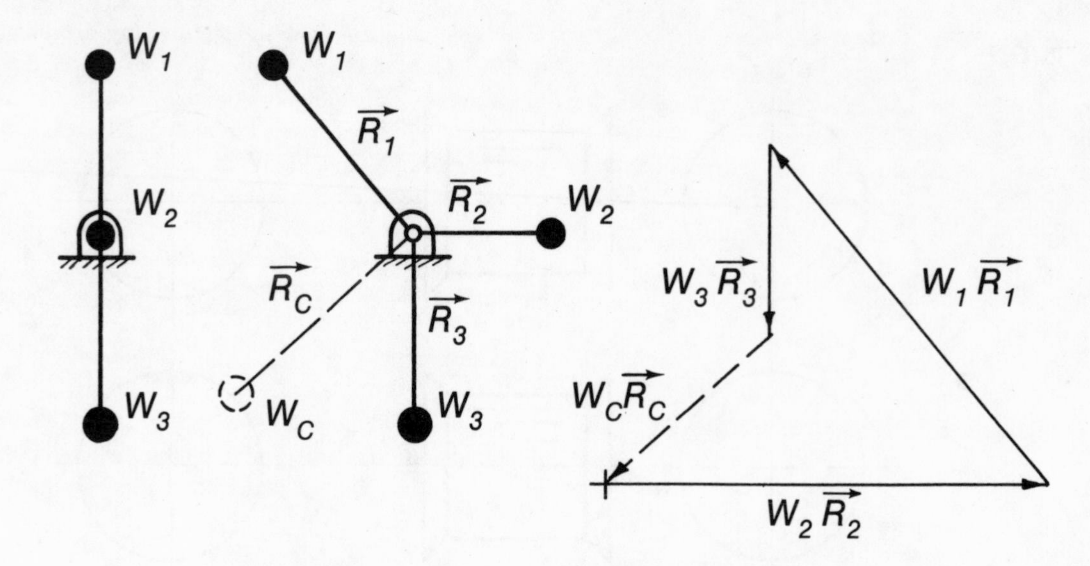

Figure 3-5. *Balancing of masses rotating in a single plane.*

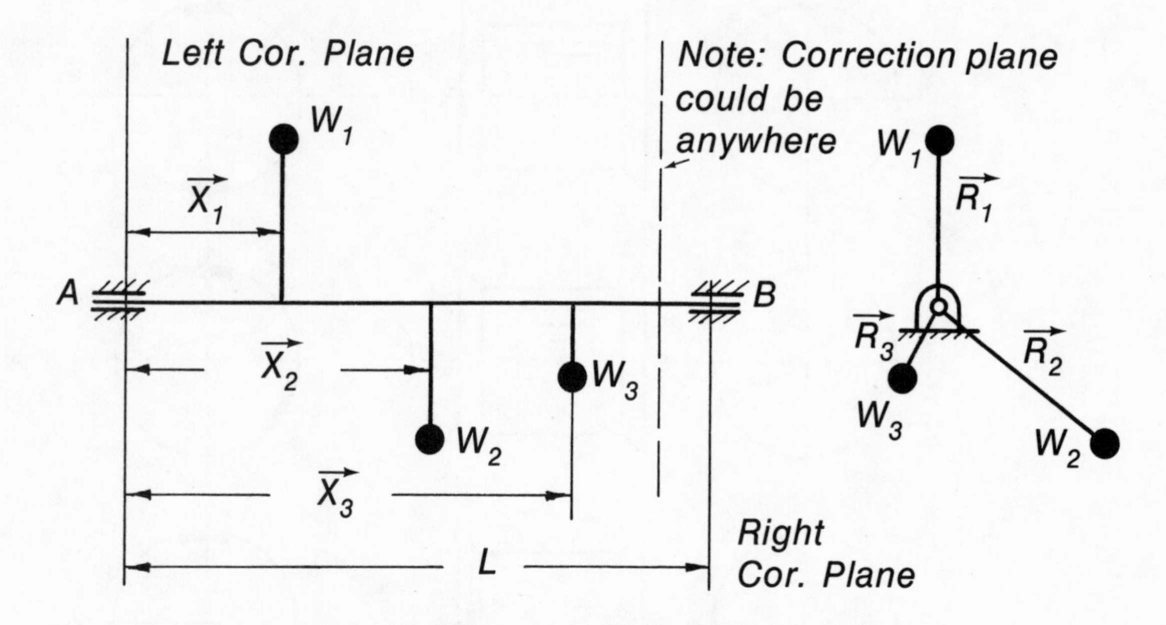

Figure 3-6. *Masses rotating in different planes.*

As an example, take a simple three-mass system rotating in a single plane as shown in Figure 3-5. You can find the magnitude and direction of the counter balance needed. The $\Sigma\vec{F} = 0$ equation could be solved using a polygon which, when solved graphically, yields the magnitude and direction of the counterbalance. This is a case of static unbalance, and all you have to do to remedy the unbalance is to apply a weight W_c at a distance R_c in the direction shown by the polygon.

You could also solve the problem analytically by writing and solving the two equations:

$$\sum_{i=1}^{3} W_i R_i \cos\theta_i = (W_c R_c)_x$$

$$\sum_{i=1}^{3} W_i R_i \sin\theta_i = (W_c R_c)_y$$

where

$$W_c R_c = \sqrt{(W_c R_c)_x^2 + (W_c R_c)_y^2}$$

and

$$\theta = \tan^{-1}\frac{(W_c R_c)_y}{(W_c R_c)_x}$$

Next, work a problem with three masses in three different planes as shown in Figure 3-6. Assume two correction planes or reference planes to solve the problem by finding the direction and magnitude of the corrections for the left and right planes. The two correction planes are chosen arbitrarily, often through the bearing supports for convenience. Do the problem in two steps.

Step 1

Take the sum of moments $\Sigma\vec{M} = 0$ about one of the correction planes or any arbitrarily chosen plane such as the one containing weight W_1, or plane A.

Using

$$\Sigma\vec{M}_A = 0; (\vec{X}_1 \times W_1\vec{R}_1) + (\vec{X}_2 \times W_2\vec{R}_2) +$$

$$(\vec{X}_3 \times W_3\vec{R}_3) + (\vec{L} \times W_B\vec{R}_B) = 0$$

or

$$W_1\vec{R}_1X_1 + W_2\vec{R}_2X_2 + W_3\vec{R}_3X_3 + W_B\vec{R}_BL = 0$$

Because all vectors in the last equation are parallel (planes are parallel), you can use a polygon to find the unknown $W_B\vec{R}_B$. Starting at 0_M, a polygon is drawn. (See Figure 3-7A.)

Determine graphically or analytically $W_B\vec{R}_B$. Draw $W_B\vec{R}_B$ on the next Figure 3-7B.

Step 2

Take summation of forces $\Sigma\vec{F} = 0$. This equation contains the terms of unknown $W_A\vec{R}_A$, known (from step 1) $W_B\vec{R}_B$, and all given $W_i\vec{R}_i$ terms. Solve it graphically or analytically, to obtain $W_A\vec{R}_A$.

When you draw a polygon $\Sigma\vec{F} = 0$, as shown in Figure 3-7B, use the equation:

$$\Sigma\vec{F} = 0; W_i\vec{R}_i + W_2\vec{R}_2 + W_3\vec{R}_3 + W_B\vec{R}_B +$$

$$W_A\vec{R}_A = 0$$

In some rotating elements, holes are drilled at convenient locations and the radius is fixed. To make the necessary correction, all you need to do is add weight, if possible, at the maximum radius (less weight will be required at maximum radius). In other situations, you can select either the radius or weight and then determine the other.

Balancing of masses reciprocating in a plane is a more complex matter. For example, it is not possible to balance a piston of a slider crank by use of a rotating mass. Only another reciprocating mass will do. In some instances, an equivalent system is possible. In other situations, dummy parts are added. Examples are pistons of engines (V-2, V-8, W, X and radial engines). Gears also are used as auxiliary arrangements for balancing.

Balancing of a four-bar linkage is shown in our flying shear mechanism example in Chapter 2, "Balancing the Shaking Force of a Four-Bar Linkage."

Example

Figure 3-8 shows a rotating shaft with three masses. Dimensions are as shown. The angular velocity is $\omega = 60$ rad/sec ccw, and constant. Calculate the bearing reactions.

Solution

$$m_1R_1\omega^2 = \frac{2}{(32.2)\ (12)}(1)\ (60)^2 = 18.63 \text{ lb} = 18.63\vec{j}$$

$$m_2R_2\omega^2 = \frac{1}{(32.2)\ (12)}(1)\ (60)^2 = 9.32 \text{ lb} =$$

$$-8.08\vec{j} + 4.66\vec{k}$$

$$m_3R_3\omega^2 = \frac{3}{(32.2)\ (12)}(.75)\ (60)^2 = 20.96 \text{ lb} =$$

$$-10.48\vec{j} - 18.15\vec{k}$$

$$\Sigma\vec{M}_A = 0;$$

$$(1\vec{i}) \times (18.63\vec{j}) + (1.75\vec{i}) \times (-8.08\vec{j} + 4.66k) +$$

$$(3\vec{i}) \times (-10.48\vec{j} - 18.15\vec{k}) + 4i \times \vec{F}_B = 0$$

$$4\vec{i} \times \vec{F}_B = \begin{vmatrix} \vec{i} & \vec{j} & \vec{k} \\ +4 & 0 & 0 \\ F_B^x & F_B^y & F_B^z \end{vmatrix} = 4F_B^y\vec{k} - 4F_B^z\vec{j}$$

$$\therefore 18.63\vec{k} - 14.12\vec{k} - 8.16\vec{j} - 31.44\vec{k} + 54.45\vec{j} +$$

$$4F_B^y\vec{k} - 4F_B^z\vec{j} = 0$$

$$-26.93\vec{k} + 46.29\vec{j} + 4F_B^y\vec{k} - 4F_B^z\vec{j} = 0$$

Comparing $\vec{j}$ and $\vec{k}$ components yields

$F_B^z = 11.57$ and $F_B^y = 6.73$

or $\vec{F}_B = 6.73\vec{j} + 11.57\vec{k} = 13.39\ \underline{/29.8°}$lb ANS.

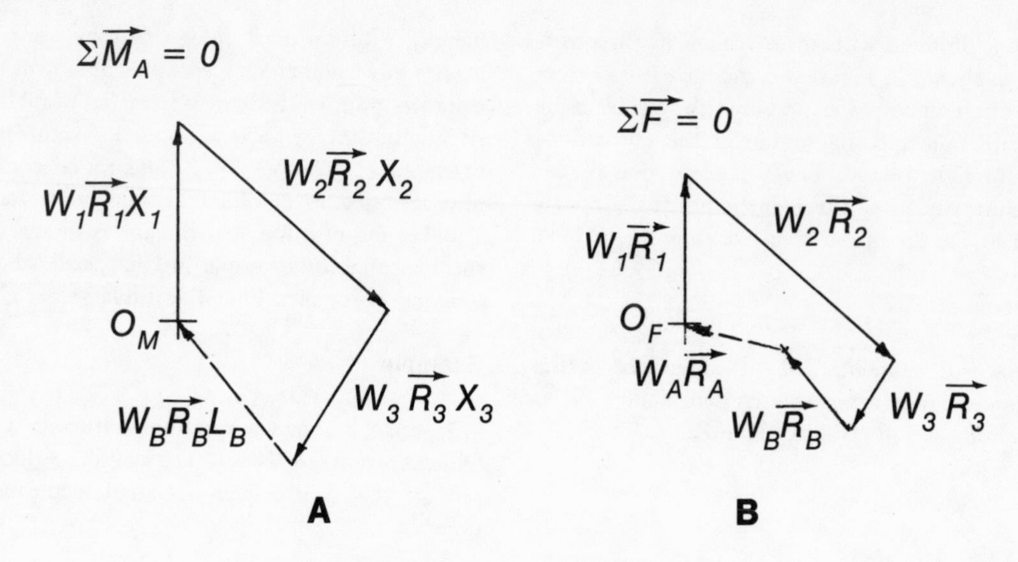

Figure 3-7. *Vector polygons.*

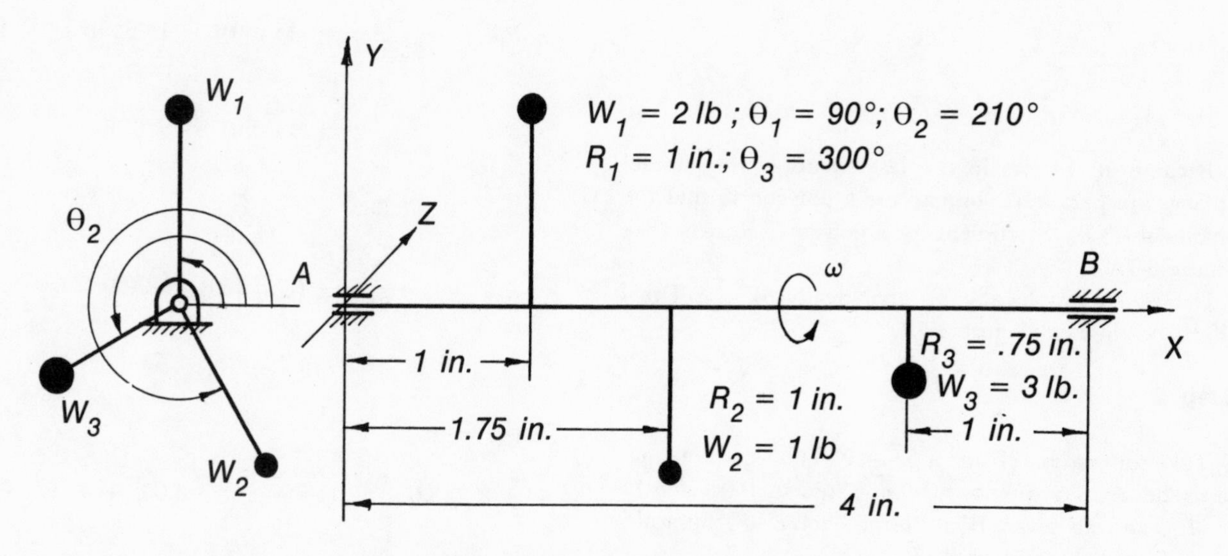

Figure 3-8. *Schematic of rotating masses.*

Similarly, taking $\Sigma\vec{M}_B = 0$ yields:

$$(-4\vec{i}) \times \vec{F}_A + (-3\vec{i}) \times (18.63\vec{j}) + (-2.25\vec{i}) \times (-8.08\vec{j} + 4.66\vec{k}) + (-1\vec{i}) \times (-10.48\vec{j} - 18.15\vec{k}) = 0$$

where

$$(-4\vec{i}) \times \vec{F}_A = -4F_A^y\vec{k} + 4F_A^z\vec{j}$$

Then:

$$-27.25\vec{k} - 7.67\vec{j} - 4F_A^y\vec{k} + 4F_A^z\vec{j} = 0$$

or

$$\vec{F}_A = -6.81\vec{j} + 1.91\vec{k} = 7.07\underline{/-74.3°} \text{ lb ANS.}$$

Review Problems

Problem 1

Find:

1. The bearing reactions at A and B for the given data and Figure 3-5
2. The magnitude and angle of the correction weight if the system is to be balanced by adding a weight located at $R_c = 10$ in. from the axis of rotation

Assume needed numerical values arbitrarily.

Given:

$\omega =$ ______ rad/sec; ccw
$W_1 =$ ______ lb
$W_2 =$ ______ lb
$W_3 =$ ______ lb
$R_1 =$ ______ in.
$R_2 =$ ______ in.
$R_3 =$ ______ in.
$\theta_1 =$ ______ deg.
$\theta_2 =$ ______ deg.
$\theta_3 =$ ______ deg.

Problem 2

Using the same data as in Problem 1, plus the following values for

$X_1 =$ ______ in.; $X_2 =$ ______ in.;
$X_3 =$ ______ in., and $l =$ ______ in.;

find the magnitude and location of the required corrections for the system shown in Figure 3-6. The system is to be balanced by using the correction planes L and R passing through the bearings A and B. Use both graphical and analytical methods. (Note: After you have finished your work, check with the partial solutions which are included in Appendix 2, solution to Chapter 3 review problem 1.)

Problem 3

Extending the number of given weights to four, five or more, and given all the data needed, write a program for a pocket calculator to solve Problem 1 and Problem 2.

References

[1] J.K. Davidson, "Programmable Calculators Take the Drudgery Out of Linkage Analysis," *Machine Design* (June 9, 1977).

[2] F.L. Conte, G.R. George, R.W. Mayne, and J.P. Sadler, "Optimum Mechanism Design Combining Kinematic and Dynamic-Force Considerations," ASME paper No. 74-DET-55.

[3] A.S. Hall, Jr., *Kinematics and Linkage Design* (Prentice-Hall, Inc., 1961).

[4] F.H. Raven, "Velocity and Acceleration Analysis of Plane and Space Mechanisms by Means of Independent-Position Equations," *Journal of Applied Mechanics, ASME,* paper No. 57-1956.

[5] M.A. Chace and P.N. Sheth, "Adaptation of Computer Techniques to the Design of Mechanical Dynamic Machinery," Transactions of ASME paper No. 73-DET-58 (1974).

Additional Reading Material

F.R.E. Crossley, *Dynamics in Machines* (The Ronald Press Company, 1954).

C.W. Ham, E.J. Crane, W.L. Rogers, *Mechanics of Machinery* (McGraw-Hill Book Company, 1958).

A.R. Holowenko, *Dynamics of Machinery* (John Wiley & Sons, Inc., 1955).

H.H. Mabie and F.W. Ocvirk, *Mechanisms and Dynamics of Machinery* (John Wiley & Sons, Inc., 1957).

G.H. Martin, *Kinematics and Dynamics of Machines* (McGraw-Hill Book Company, 1969).

R.L. Maxwell, *Kinematics and Dynamics of Machinery* (Prentice-Hall, Inc., 1960).

B. Paul, "Analytical Dynamics of Mechanisms—A Computer Oriented Overview," *Mechanisms and Machine Theory,* Vol. 10 (1975).

R.M. Phelan, *Dynamics of Machinery* (McGraw-Hill Book Company, 1967).

J.E. Shigley, *Theory of Machines* (McGraw-Hill Book Company, 1961).

C.H. Suh, *Kinematics and Mechanisms Design* (John Wiley & Sons, Inc., 1978).

Appendix 1

TI-59 Program

This program uses complex numbers in development of the equations needed to determine velocities, accelerations, forces, and torques. Complex numbers are particularly convenient in describing links in mechanisms. As an example, the magnitude of the complex number is the link length, while the angle represents the direction of the link.

There are two ways of writing complex numbers—rectangular and exponential-polar forms. Conversions from one form to the other are easy by using ready-programmed subroutines available in most pocket calculators.

The following formulas are useful in linkage calculations. Rectangular forms:

$$r = x + jy$$

$$r = |r|\ (\cos\theta + j\sin\theta)$$

Polar forms:

$$r = |r|\ \underline{/\theta}$$

$$r = |r|\ e^{j\theta}$$

Conversion formulas:

$$|r| = \sqrt{x^2 + y^2}; \qquad \theta = \tan^{-1}\frac{y}{x}$$

$$x = |r| \cos\theta\ ; \qquad y = |r| \sin\theta$$

In order to use the program on a TI-59 calculator, the following steps need to be taken:

1. Familiarize yourself with the owner's manual for the TI Programmable 58/59 to make sure that proper procedures are used—how to make corrections in a program, how to partition from the original 479.59 to the required 639.39 steps needed (press 4 OP 17), and how to record on magnetic cards (press INV.FIX, 1,2,3,4, WRITE; two cards are required), and read them (press 1,2,3 or 4 and insert the corresponding card).
2. After pressing the LRN instruction, record all, or parts, of the program as needed. If only the kinematics part is needed, the program could be terminated by adding an R/S instruction immediately after the last instruction of the angular acceleration portion of the whole program has been recorded.
3. When using the printer PC 100 A with the calculator, by pressing the LIST instruction from the keyboard the whole program will be printed and could be easily checked for possible mistakes in recording. Also, the memory could be checked to make sure that the data were stored as given.
4. Finally, to make sure that the program has been recorded correctly, a trial run could be made using the example given at the end of the program included in this appendix. The results obtained should be the same as those printed in the example. It takes about one minute to run the entire program. Comment: There is no doubt that this program could be improved and shortened.

Two cases need to be considered for a four-bar linkage. They are the open and crossed types, as shown by the solid lines and the broken lines respectively, in Figure A. The only difference will be in determining the angles θ_3 and θ_4 for the two types.

For both types of linkage, the law of cosines and other relations are

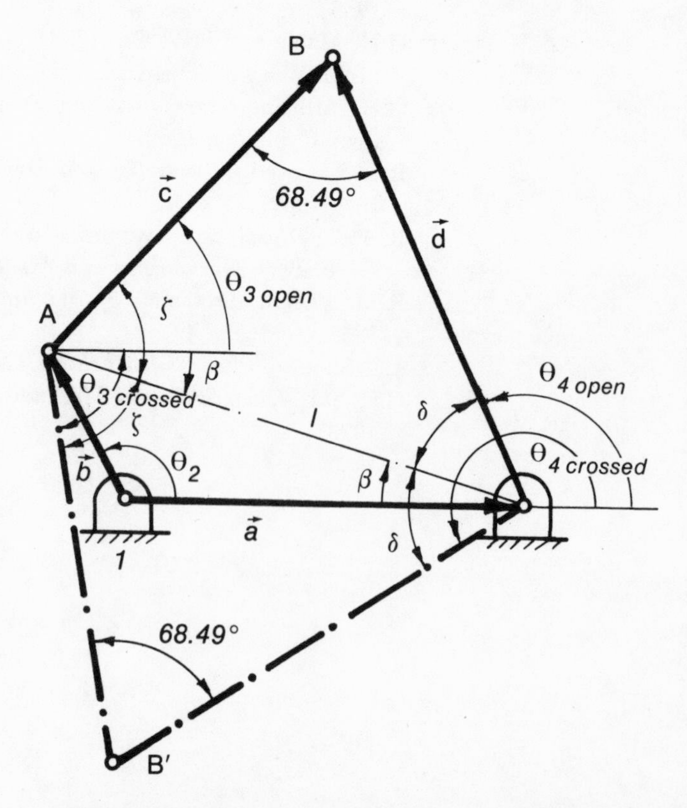

Figure A. Open and crossed four-bar linkage.

$$l^2 = a^2 + b^2 - 2ab \cos (a,b)$$

$$\tan \beta = b \sin \theta_2/(a - b \cos \theta_2)$$

$$c^2 = d^2 + l^2 - 2dl \cos \delta$$

$$d^2 = c^2 + l^2 - 2cl \cos \gamma$$

Starting with the open linkage, where from Figure A

$$\theta_3 = \gamma - \beta$$

$$\theta_4 = 180 - \delta - \beta$$

and for the crossed linkage,

$$\theta_3 = -\gamma - \beta$$

$$\theta_4 = 180 + \delta - \beta$$

These formulas will be used for the determination of the necessary values for θ_3 and θ_4 of this problem.

For example, if $\theta_2 = 120°$, $a = 10$ in., $b = 4$ in., $c = 10$ in. and $d = 12$ in., the following will result:

$l = 12.49$; $\beta = 16.1°$; $\gamma = 63.4°$; $\delta = 48.1°$ and
$\theta_3 = 47.3°$; $\theta_3 = -79.5°$; $\theta_4 = 115.7°$ and $\theta_4 = 227°$

for both the open and crossed linkages.

In order to obtain the velocity and acceleration equations for this problem, let the four links be represented by vectors $\vec{a}, \vec{b}, \vec{c},$ and $\vec{d}$. Then you can write:

$$\vec{a} - \vec{b} - \vec{c} + \vec{d} = 0$$

or

$$a\, e^{j\theta_1} - b\, e^{j\theta_2} - c\, e^{j\theta_3} + d\, e^{j\theta_4} = 0$$

The velocities and accelerations for different points can be obtained easily once the angular velocities and angular accelerations for link c and link d are determined. The above equation will be differentiated as follows:

$$a\, e^{j\theta_1} \frac{d\theta_1}{dt} - j\, b\, e^{j\theta_2} \frac{d\theta_2}{dt} + j\, c\, e^{j\theta_3} \frac{d\theta_3}{dt} +$$

$$j\, d\, e^{j\theta_4} \frac{d\theta_4}{dt} = 0$$

or

$$-j\, b\omega_2 (\cos \theta_2 + j \sin \theta_2) - j\, c\omega_3 (\cos \theta_3 + j \sin \theta_3) +$$

$$j\, d\omega_4 (\cos \theta_4 + j \sin \theta_4) = 0$$

Separating real and imaginary parts yields the following system of two equations:

$$\left.\begin{aligned} -c \cos \theta_3\omega_3 + d \cos \theta_4\omega_4 &= b \cos \theta_2\omega_2 \\ -c \sin \theta_3\omega_3 + d \sin \theta_4\omega_4 &= b \sin \theta_2\omega_2 \end{aligned}\right\}$$

Using Cramer's rule and the trigonometric relations

$$\sin (\alpha - \beta) = \sin \alpha \cos \beta - \cos \alpha \sin \beta$$

You obtain the formulas for ω_3 and ω_4. They are:

$$\omega_3 = \frac{b\omega_2 \cos \theta_2 d \sin \theta_4 - b\omega_2 \sin \theta_2 d \cos \theta_4}{c \cos \theta_3 d \sin \theta_4 + c \sin \theta_3 d \cos \theta_4}$$

$$\omega_4 = \text{(similar)}$$

or

$$\omega_3 = \frac{-b\omega_2 \sin (\theta_2 - \theta_4)}{c \sin (\theta_3 - \theta_4)}$$

and

$$\omega_4 = \frac{b\omega_2 \sin (\theta_2 - \theta_3)}{d \sin (\theta_4 - \theta_3)}$$

To obtain the angular accelerations you differentiate eq. (1) once more:

$$-\frac{db}{dt} e^{j\theta_2} j \frac{d\theta_2}{dt} - b\, e^{j\theta_2} j \frac{d\theta_2}{dt} j \frac{d\theta_2}{dt} - b\, e^{j\theta_2} j \frac{d^2\theta_2}{dt^2} -$$

$$\frac{dc}{dt} e^{j\theta_3} j \frac{d\theta_3}{dt} - c\, e^{j\theta_3} j \frac{d\theta_3}{dt} j \frac{d\theta_3}{dt} - c\, e^{j\theta_3} j \frac{d^2\theta_3}{dt^2} +$$

$$\frac{dd}{dt} e^{j\theta_4} j \frac{d\theta_4}{dt} + d\, e^{j\theta_4} j \frac{d\theta_4}{dt} j \frac{d\theta_4}{dt} + d\, e^{j\theta_4} j \frac{d^2\theta_4}{dt^2} = 0$$

Eliminating zero terms

$$\frac{db}{dt} = \frac{dc}{dt} = \frac{dd}{dt} = 0$$

and identifying new terms

$$\frac{d\theta}{dt} = \omega; \frac{d^2\theta}{dt^2} = \alpha; \text{ and } j^2 = -1$$

yields

$$b\, e^{j\theta_2} \omega_2^2 - jb\, e^{j\theta_2} \alpha_2$$

$$+ c\, e^{j\theta_3} \omega_3^2 - jc\, e^{j\theta_3} \alpha_3$$

$$- d\, e^{j\theta_4} \omega_4^2 + jd\, e^{j\theta_4} \alpha_4 = 0$$

Converting from $e^{j\theta}$ form to rectangular form

$$b\,\omega_2^2(\cos\theta_2 + j\sin\theta_2) - j\,b\,\alpha_2(\cos\theta_2 + j\sin\theta_2) +$$

$$c\,\omega_3^2(\cos\theta_3 + j\sin\theta_3) - j\,c\,\alpha_3(\cos\theta_3 + j\sin\theta_3) -$$

$$d\,\omega_4^2(\cos\theta_4 + j\sin\theta_4) + j\,d\,\alpha_4(\cos\theta_4 + j\sin\theta_4) = 0$$

Again, separating into real and imaginary parts you obtain:

$$c\sin\theta_3\alpha_3 - d\sin\theta_4\alpha_4 = -A$$

$$-c\cos\theta_3\alpha_3 + d\cos\theta_4\alpha_4 = -B$$

where:

$$A = b\,\omega_2^2\cos\theta_2 + b\,\alpha_2\sin\theta_2 + c\,\omega_3^2\cos\theta_3 - d\,\omega_4^2\cos\theta_4$$

$$B = b\,\omega_2^2\sin\theta_2 - b\,\alpha_2\cos\theta_2 + c\,\omega_3^2\sin\theta_3 - d\,\omega_4^2\sin\theta_4$$

using Cramer's rule you solve the system of the two equations for α_3 and α_4. Then,

$$\alpha_3 = [d\,\omega_4^2(\cos^2\theta_4 + \sin^2\theta_4) - b\,\omega_2^2(\sin\theta_2\sin\theta_4 +$$

$$\cos\theta_2\cos\theta_4) - b\,\alpha_2(\sin\theta_2\cos\theta_4 - \cos\theta_2\sin\theta_4) -$$

$$c\,\omega_3^2(\cos\theta_3\cos\theta_4 + \sin\theta_3\sin\theta_4)]/cd\sin(\theta_3 - \theta_4)$$

Substituting,

$$\sin^2\alpha + \cos^2\alpha = 1$$

$$\sin\alpha\sin\beta + \cos\alpha\cos\beta = \cos(\alpha - \beta)$$

$$\sin\alpha\sin\beta - \cos\alpha\sin\beta = \sin(\alpha - \beta)$$

$$\sin\alpha\cos\beta - \cos\alpha\sin\beta = \sin(\alpha - \beta)$$

You obtain α_3 as shown. Similarly for α_4 (after repeating the procedure), you obtain:

$$\alpha_3 = \frac{d\omega_4^2 - b\omega_2^2\cos(\theta_2 - \theta_4) - b\alpha_2\sin(\theta_2 - \theta_4) - c\omega_3^2\cos(\theta_3 - \theta_4)}{c\sin(\theta_3 - \theta_4)}$$

$$\alpha_4 = \frac{c\omega_3^2 + b\omega_2^2\cos(\theta_2 - \theta_3) + b\alpha_2\sin(\theta_2 - \theta_3) - d\omega_4^2\cos(\theta_4 - \theta_3)}{d\sin(\theta_4 - \theta_3)}$$

The equations now will be used to write the program. A step-by-step procedure can be seen by merely copying the program and realizing that the following storage for the terms in the previously given equations were used.

STO	Description
00	
01	Frame link length a (in, ft)
02	Crank link length b
03	Coupler length c
04	Driven link length d
05	θ_2 in degrees; positive if ccw
06	ω_2 rad/sec; positive if ccw
07	ω_3 open; rad/sec, if ccw positive
08	ω_3 crossed, rad/sec, also W_3 weight, lb.
09	ω_4 open, rad/sec
10	ω_4 crossed, rad/sec
11	α_3 open; rad/sec^2; if ccw positive
12	α_3 crossed, rad/sec^2, also $\bar{I}_3$, moment of inertia
13	α_4 open, rad/sec^2
14	α_4 crossed, rad/sec^2, also mA_{G3}—inertia force, lb.
15	θ_3 open in degrees, positive if ccw
16	θ_3 crossed in degrees, also h_3, distance, in.
17	θ_4 crossed in degrees
18	θ_4 open in degrees
19	γ angle in degrees
20	β angle in degrees
21	c sin $(\theta_3 - \theta_4)$ open
22	c sin $(\theta_3 - \theta_4)$ crossed, also γ_3, angle, degree
23	d sin$(\theta_4 - \theta_3)$ open
24	d sin $(\theta_4 - \theta_3)$ crossed
25	α_2 rad/sec^2
26	l in.; ft
27	angle δ, degrees
28	rg_3 - location of gravity center of link 3, in.; ft
29	Real component of A_{G3}
30	Imaginary component of A_{G3}
31	Value of A_{G3} - Accel. of gravity center
32	Angle of inclination of the acceleration of gravity center with horizontal.
33	r, distance of m_3A_{G3} from pt. A
34	T_2, torque, in.-lb, ft-lb
35	
36	F_{23}, force due to inertia of link 3 (lb)
37	F_{43}, force due to inertia of link 3
38	F_{23} real
39	F_{23} imaginary
40	

Description of the Program

The first program denoted as "Kinematics and Dynamics of an Open Four-Bar Linkage: 639.39" has been made of two parts—kinematics and dynamics. The kinematics part ends at step 464-95, and the dynamics part ends with step 630-91. The dynamics formulas are given in the section "Force Analysis Using Complex Numbers Method," Chapter 2. The kinematics part formulas are given in this appendix. The step-by-step procedure follows.

Kinematics

000-150 geometry of the linkage
154-192 calculations of ω_3
197-234 calculations for ω_4
239-299 calculations for α_3
303-363 calculations for α_4
367-464 calculations for A_{G3}

Dynamics

468-540 calculations of F'_{43}
542-565 calculation of F'_{23} real
566-589 calculation of F'_{23} imaginary
590-601 calculation of F'_{23}
602-612 calculations of γ_3
613-630 torque calculations T_2

The second program, "Kinematics of a Crossed Four-Bar Linkage: 479.59," is a step-by-step procedure for calculating the following:

000-150 geometry of linkage
151-193 angular velocity of link 3, ω_3
194-235 angular velocity of link 4, ω_4
236-299 angular acceleration of link 3, α_3
300-363 angular acceleration of link 4, α_4
364-464 acceleration of center of gravity of link 3; A_{G3}

(Note: Some of the memory of the crossed linkage has been used in the dynamics part of the open linkage as shown in the descriptions.)

Kinematics and Dynamics of an Open Four-Bar Linkage: 639.39

Step	Code	Key Entry
000	76	LBL
001	11	A
002	58	FIX
003	04	04
004	53	(
005	53	(
006	53	(
007	43	RCL
008	01	01
009	33	X^2
010	85	+
011	43	RCL
012	02	02
013	33	X^2
014	75	-
015	02	2
016	65	×
017	43	RCL
018	01	01
019	65	×
020	43	RCL
021	02	02
022	65	×
023	43	RCL
024	05	05
025	39	COS
026	54	)
027	42	STO
028	26	26
029	85	+
030	43	RCL
031	03	03
032	33	X^2
033	75	-
034	43	RCL
035	04	04
036	33	X^2
037	54	)
038	55	÷
039	02	2
040	55	÷
041	43	RCL
042	03	03
043	55	÷
044	43	RCL
045	26	26
046	34	√X
047	54	)
048	22	INV
049	39	COS
050	42	STO
051	19	19
052	53	(
053	53	(
054	43	RCL
055	04	04
056	33	X^2
057	75	-
058	43	RCL
059	03	03
060	33	X^2
061	85	+
062	43	RCL
063	26	26
064	54	)
065	55	÷
066	02	2
067	55	÷
068	01	1
069	55	÷
070	43	RCL
071	04	04
072	55	÷
073	43	RCL
074	26	26
075	34	√X
076	54	)
077	22	INV
078	39	COS
079	42	STO
080	27	27
081	53	(
082	43	RCL
083	02	02
084	65	×
085	43	RCL
086	05	05
087	38	SIN
088	55	÷
089	53	(
090	43	RCL
091	01	01
092	75	-
093	43	RCL

Step	Code	Key Entry
094	02	02
095	65	×
096	43	RCL
097	05	05
098	39	COS
099	54	)
100	54	)
101	22	INV
102	30	TAN
103	42	STO
104	20	20
105	53	(
106	43	RCL
107	19	19
108	75	-
109	43	RCL
110	20	20
111	54	)
112	42	STO
113	15	15
114	53	(
115	43	RCL
116	19	19
117	94	+/-
118	75	-
119	43	RCL
120	20	20
121	54	)
122	42	STO
123	16	16
124	53	(
125	01	1
126	08	8
127	00	0
128	75	-
129	43	RCL
130	27	27
131	75	-
132	43	RCL
133	20	20
134	54	)
135	42	STO
136	18	18
137	53	(
138	01	1
139	08	8
140	00	0
141	85	+
142	43	RCL
143	27	27
144	75	-
145	43	RCL
146	20	20
147	54	)
148	42	STO
149	17	17
150	95	=
151	91	R/S
152	76	LBL
153	12	B
154	53	(
155	53	(
156	43	RCL
157	02	02
158	65	×
159	43	RCL
160	06	06
161	65	×
162	53	(
163	43	RCL
164	05	05
165	75	-
166	43	RCL
167	18	18
168	54	)
169	38	SIN
170	54	)
171	55	÷
172	53	(
173	53	(
174	43	RCL
175	03	03
176	65	×
177	53	(
178	43	RCL
179	15	15
180	75	-
181	43	RCL
182	18	18
183	54	)
184	38	SIN
185	54	)
186	42	STO
187	21	21
188	54	)
189	94	+/-
190	54	)
191	42	STO
192	07	07
193	95	=
194	91	R/S
195	76	LBL
196	13	C
197	53	(
198	53	(
199	43	RCL
200	02	02
201	65	×
202	43	RCL
203	06	06
204	65	×
205	53	(
206	43	RCL
207	05	05
208	75	-
209	43	RCL
210	15	15
211	54	)
212	38	SIN
213	54	)
214	55	÷
215	53	(
216	53	(
217	43	RCL
218	04	04
219	65	×
220	53	(
221	43	RCL
222	18	18
223	75	-
224	43	RCL
225	15	15
226	54	)
227	38	SIN
228	54	)
229	42	STO
230	23	23
231	54	)
232	54	)
233	42	STO
234	09	09
235	95	=
236	91	R/S
237	76	LBL
238	14	D
239	53	(
240	43	RCL
241	04	04
242	65	×
243	43	RCL
244	09	09
245	33	X^2
246	75	-
247	43	RCL
248	02	02
249	65	×
250	43	RCL
251	06	06
252	33	X^2
253	65	×
254	53	(
255	43	RCL
256	05	05
257	75	-
258	43	RCL
259	18	18
260	54	)
261	39	COS
262	75	-
263	43	RCL
264	02	02
265	65	×
266	43	RCL
267	25	25
268	65	×
269	53	(
270	43	RCL
271	05	05
272	75	-
273	43	RCL
274	18	18
275	54	)
276	38	SIN
277	75	-
278	43	RCL
279	03	03
280	65	×
281	43	RCL
282	07	07
283	33	X^2
284	65	×
285	53	(
286	43	RCL
287	15	15
288	75	-
289	43	RCL
290	18	18
291	54	)
292	39	COS
293	54	)
294	55	÷
295	43	RCL
296	21	21

Step	Code	Key Entry
297	95	=
298	42	STO
299	11	11
300	91	R/S
301	76	LBL
302	15	E
303	53	(
304	43	RCL
305	03	03
306	65	×
307	43	RCL
308	07	07
309	33	X^2
310	85	+
311	43	RCL
312	02	02
313	65	×
314	43	RCL
315	06	06
316	33	X^2
317	65	×
318	53	(
319	43	RCL
320	05	05
321	75	-
322	43	RCL
323	15	15
324	54	)
325	39	COS
326	85	+
327	43	RCL
328	02	02
329	65	×
330	43	RCL
331	25	25
332	65	×
333	53	(
334	43	RCL
335	05	05
336	75	-
337	43	RCL
338	15	15
339	54	)
340	38	SIN
341	75	-
342	43	RCL
343	04	04
344	65	×
345	43	RCL
346	09	09
347	33	X^2
348	65	×
349	53	(
350	43	RCL
351	18	18
352	75	-
353	43	RCL
354	15	15
355	54	)
356	39	COS
357	54	)
358	55	÷
359	43	RCL
360	23	23
361	95	=
362	42	STO
363	13	13
364	91	R/S
365	76	LBL
366	16	A'
367	53	(
368	43	RCL
369	02	02
370	94	+/-
371	65	×
372	43	RCL
373	06	06
374	33	X^2
375	65	×
376	43	RCL
377	05	05
378	39	COS
379	75	-
380	43	RCL
381	28	28
382	65	×
383	43	RCL
384	11	11
385	65	×
386	43	RCL
387	15	15
388	38	SIN
389	75	-
390	43	RCL
391	28	28
392	65	×
393	43	RCL
394	07	07
395	33	X^2
396	65	×
397	43	RCL
398	15	15
399	39	COS
400	54	)
401	42	STO
402	29	29
403	53	(
404	43	RCL
405	02	02
406	94	+/-
407	65	×
408	43	RCL
409	06	06
410	33	X^2
411	65	×
412	43	RCL
413	05	05
414	38	SIN
415	85	+
416	43	RCL
417	28	28
418	65	×
419	43	RCL
420	11	11
421	65	×
422	43	RCL
423	15	15
424	39	COS
425	75	-
426	43	RCL
427	28	28
428	65	×
429	43	RCL
430	07	07
431	33	X^2
432	65	×
433	43	RCL
434	15	15
435	38	SIN
436	54	)
437	42	STO
438	30	30
439	53	(
440	43	RCL
441	29	29
442	32	X:T
443	43	RCL
444	30	30
445	22	INV
446	37	P/R
447	42	STO
448	32	32
449	32	X:T
450	42	STO
451	31	31
452	54	)
453	95	=
454	66	PAU
455	66	PAU
456	66	PAU
457	66	PAU
458	66	PAU
459	66	PAU
460	66	PAU
461	66	PAU
462	66	PAU
463	66	PAU
464	66	PAU
465	66	PAU
466	76	LBL
467	17	B'
468	53	(
469	43	RCL
470	12	12
471	65	×
472	43	RCL
473	11	11
474	55	÷
475	53	(
476	43	RCL
477	08	08
478	55	÷
479	03	3
480	08	8
481	06	6
482	65	×
483	43	RCL
484	31	31
485	54	)
486	42	STO
487	14	14
488	54	)
489	42	STO
490	16	16
491	53	(
492	43	RCL
493	28	28
494	85	+
495	53	(
496	43	RCL
497	16	16
498	55	÷
499	53	(
500	43	RCL

Step	Code	Key Entry
501	32	32
502	75	-
503	43	RCL
504	15	15
505	54	)
506	38	SIN
507	54	)
508	54	)
509	42	STO
510	33	33
511	53	(
512	53	(
513	43	RCL
514	32	32
515	75	-
516	43	RCL
517	15	15
518	54	)
519	38	SIN
520	55	÷
521	53	(
522	43	RCL
523	18	18
524	75	-
525	43	RCL
526	15	15
527	54	)
528	38	SIN
529	65	×
530	43	RCL
531	14	14
532	65	×
533	43	RCL
534	33	33
535	55	÷
536	43	RCL
537	03	03
538	54	)
539	42	STO
540	37	37
541	37	P/R
542	53	(
543	43	RCL
544	37	37
545	94	+/-
546	65	×
547	43	RCL
548	18	18
549	39	COS
550	75	-
551	43	RCL
552	14	14
553	65	×
554	53	(
555	43	RCL
556	32	32
557	85	+
558	01	1
559	08	8
560	00	0
561	54	)
562	39	COS
563	54	)
564	42	STO
565	38	38
566	53	(
567	43	RCL
568	37	37
569	94	+/-
570	65	×
571	43	RCL
572	18	18
573	38	SIN
574	75	-
575	43	RCL
576	14	14
577	65	×
578	53	(
579	43	RCL
580	32	32
581	85	+
582	01	1
583	08	8
584	00	0
585	54	)
586	38	SIN
587	54	)
588	42	STO
589	39	39
590	53	(
591	43	RCL
592	38	38
593	32	X⇌T
594	43	RCL
595	39	39
596	22	INV
597	37	P/R
598	42	STO
599	22	22
600	32	X⇌T
601	42	STO
602	36	36
603	54	)
604	65	×
605	43	RCL
606	02	02
607	65	×
608	53	(
609	43	RCL
610	05	05
611	75	-
612	43	RCL
613	22	22
614	54	)
615	38	SIN
616	95	=
617	94	+/-
618	42	STO
619	34	34
620	66	PAU
621	66	PAU
622	66	PAU
623	66	PAU
624	00	0
625	22	INV
626	90	LST
627	68	NOP
628	68	NOP
629	68	NOP
630	91	R/S

Value	Register
0.	00
7.	01
3.	02
8.	03
6.	04
150.	05
200.	06
42.20372884	07
0.7	08
86.02907652	09
0.	10
6975.257828	11
0.02	12
-8319.059236	13
168.1982906	14
29.18342038	15
.8294088842	16
226.4120123	17
115.8231661	18
38.06583114	19
8.882410764	20
-7.986245826	21
-10.94309779	22
5.989684369	23
0.	24
0.	25
94.37306696	26
55.29442309	27
4.	28
84098.06643	29
-39114.65033	30
92749.34312	31
-24.94349671	32
2.976439024	33
-130.0781007	34
0.	35
132.7976569	36
-50.79615697	37
130.3828906	38
-25.20951267	39

Kinematics of a Crossed Four-Bar Linkage: 479.59

Step	Code	Key Entry
000	76	LBL
001	11	A
002	58	FIX
003	04	04
004	53	(
005	53	(
006	53	(
007	43	RCL
008	01	01
009	33	X²
010	85	+
011	43	RCL
012	02	02
013	33	X²
014	75	-
015	02	2
016	65	×
017	43	RCL
018	01	01
019	65	×
020	43	RCL
021	02	02
022	65	×
023	43	RCL
024	05	05
025	39	COS
026	54	)
027	42	STO
028	26	26
029	85	+
030	43	RCL
031	03	03
032	33	X²
033	75	-
034	43	RCL
035	04	04
036	33	X²
037	54	)
038	55	÷
039	02	2
040	55	÷
041	43	RCL
042	03	03
043	55	÷
044	43	RCL
045	26	26
046	34	√X
047	54	)
048	22	INV
049	39	COS
050	42	STO
051	19	19
052	53	(
053	53	(
054	43	RCL
055	04	04
056	33	X²
057	75	-
058	43	RCL
059	03	03
060	33	X²
061	85	+
062	43	RCL
063	26	26
064	54	)
065	55	÷
066	02	2
067	55	÷
068	01	1
069	55	÷
070	43	RCL
071	04	04
072	55	÷
073	43	RCL
074	26	26
075	34	√X
076	54	)
077	22	INV
078	39	COS
079	42	STO
080	27	27
081	53	(
082	43	RCL
083	02	02
084	65	×
085	43	RCL
086	05	05
087	38	SIN
088	55	÷
089	53	(
090	43	RCL
091	01	01
092	75	-
093	43	RCL
094	02	02
095	65	×
096	43	RCL
097	05	05
098	39	COS
099	54	)
100	54	)
101	22	INV
102	30	TAN
103	42	STO
104	20	20
105	53	(
106	43	RCL
107	19	19
108	75	-
109	43	RCL
110	20	20
111	54	)
112	42	STO
113	15	15
114	53	(
115	43	RCL
116	19	19
117	94	+/-
118	75	-
119	43	RCL
120	20	20
121	54	)
122	42	STO
123	16	16
124	53	(
125	01	1
126	08	8
127	00	0
128	75	-
129	43	RCL
130	27	27
131	75	-
132	43	RCL
133	20	20
134	54	)
135	42	STO
136	18	18
137	53	(
138	01	1
139	08	8
140	00	0
141	85	+
142	43	RCL
143	27	27
144	75	-
145	43	RCL
146	20	20
147	54	)
148	42	STO
149	17	17
150	95	=
151	76	LBL
152	17	B'
153	53	(
154	53	(
155	43	RCL
156	02	02
157	65	×
158	43	RCL
159	06	06
160	65	×
161	53	(
162	43	RCL
163	05	05
164	75	-
165	43	RCL
166	17	17
167	54	)
168	38	SIN
169	54	)
170	55	÷
171	53	(
172	53	(
173	43	RCL
174	03	03
175	65	×
176	53	(
177	43	RCL
178	16	16
179	75	-
180	43	RCL
181	17	17
182	54	)
183	38	SIN
184	54	)
185	42	STO
186	22	22
187	54	)
188	94	+/-
189	54	)
190	42	STO
191	08	08
192	95	=
193	91	R/S
194	76	LBL
195	18	C'

Step	Code	Key Entry
196	53	(
197	53	(
198	43	RCL
199	02	02
200	65	×
201	43	RCL
202	06	06
203	65	×
204	53	(
205	43	RCL
206	05	05
207	75	-
208	43	RCL
209	16	16
210	54	)
211	38	SIN
212	54	)
213	55	÷
214	53	(
215	53	(
216	43	RCL
217	04	04
218	65	×
219	53	(
220	43	RCL
221	17	17
222	75	-
223	43	RCL
224	16	16
225	54	)
226	38	SIN
227	54	)
228	42	STO
229	24	24
230	54	)
231	54	)
232	42	STO
233	10	10
234	95	=
235	91	R/S
236	76	LBL
237	19	D'
238	53	(
239	43	RCL
240	04	04
241	65	×
242	43	RCL
243	10	10
244	33	X^2
245	75	-
246	43	RCL
247	02	02
248	65	×
249	43	RCL
250	06	06
251	33	X^2
252	65	×
253	53	(
254	43	RCL
255	05	05
256	75	-
257	43	RCL
258	17	17
259	54	)
260	39	COS
261	75	-
262	43	RCL
263	02	02
264	65	×
265	43	RCL
266	25	25
267	65	×
268	53	(
269	43	RCL
270	05	05
271	75	-
272	43	RCL
273	17	17
274	54	)
275	38	SIN
276	75	-
277	43	RCL
278	03	03
279	65	×
280	43	RCL
281	08	08
282	33	X^2
283	65	×
284	53	(
285	43	RCL
286	16	16
287	75	-
288	43	RCL
289	17	17
290	54	)
291	39	COS
292	54	)
293	55	÷
294	43	RCL
295	22	22
296	95	=
297	42	STO
298	12	12
299	91	R/S
300	76	LBL
301	10	E'
302	53	(
303	43	RCL
304	03	03
305	65	×
306	43	RCL
307	08	08
308	33	X^2
309	85	+
310	43	RCL
311	02	02
312	65	×
313	43	RCL
314	06	06
315	33	X^2
316	65	×
317	53	(
318	43	RCL
319	05	05
320	75	-
321	43	RCL
322	16	16
323	54	)
324	39	COS
325	85	+
326	43	RCL
327	02	02
328	65	×
329	43	RCL
330	25	25
331	65	×
332	53	(
333	43	RCL
334	05	05
335	75	-
336	43	RCL
337	16	16
338	54	)
339	38	SIN
340	75	-
341	43	RCL
342	04	04
343	65	×
344	43	RCL
345	10	10
346	33	X^2
347	65	×
348	53	(
349	43	RCL
350	17	17
351	75	-
352	43	RCL
353	16	16
354	54	)
355	39	COS
356	54	)
357	55	÷
358	43	RCL
359	24	24
360	95	=
361	42	STO
362	14	14
363	91	R/S
364	76	LBL
365	16	A'
366	53	(
367	43	RCL
368	02	02
369	94	+/-
370	65	×
371	43	RCL
372	06	06
373	33	X^2
374	65	×
375	43	RCL
376	05	05
377	39	COS
378	75	-
379	43	RCL
380	28	28
381	65	×
382	43	RCL
383	12	12
384	65	×
385	43	RCL
386	16	16
387	38	SIN
388	75	-
389	43	RCL
390	28	28
391	65	×
392	43	RCL
393	08	08
394	33	X^2
395	65	×
396	43	RCL
397	16	16
398	39	COS

Step	Key Code	Key Entry
399	54	)
400	42	STO
401	29	29
402	53	(
403	43	RCL
404	02	02
405	94	+/-
406	65	×
407	43	RCL
408	06	06
409	33	X^2
410	65	×
411	43	RCL
412	05	05
413	38	SIN
414	85	+
415	43	RCL
416	28	28
417	65	×
418	43	RCL
419	12	12
420	65	×
421	43	RCL
422	16	16
423	39	COS
424	75	-
425	43	RCL
426	28	28
427	65	×
428	43	RCL
429	08	08
430	33	X^2
431	65	×
432	43	RCL
433	16	16
434	38	SIN
435	54	)
436	42	STO
437	30	30
438	53	(
439	53	(
440	43	RCL
441	29	29
442	33	X^2
443	85	+
444	43	RCL
445	30	30
446	33	X^2
447	54	)
448	34	√X
449	42	STO
450	31	31
451	54	)
452	53	(
453	43	RCL
454	30	30
455	55	÷
456	43	RCL
457	29	29
458	54	)
459	22	INV
460	30	TAN
461	42	STO
462	32	32
463	95	=
464	91	R/S

The HP-67, HP-97 Calculator Programs

The Hewlett-Packard Company provides for its HP-67 and HP-97 calculators a "M.E. Pac 1" with 23 programs of which few are of interest here. The HP-67 pocket calculator has 224 steps of programming capacity. The programs are stored on magnetic cards. The first two programs are related to function generation and progression of a four-bar system. The third program handles the progression of a slider-crank mechanism. The fourth program computes the reaction forces due to a torque applied to gears (helical, bevel, and worm gears). The progression of a four-bar system calculates in a manual or automatic mode the displacements, angular velocities and accelerations of the output link whenever the input angle and angular velocity and acceleration are given. Basically, it uses the same equations for determination of the angles of the coupler and output links as included in the program. The angular velocities and accelerations are obtained next by differentiating the previously calculated angles of the particular linkage.

Of particular interest is the programs' capacity of automatically progressing from a given input angle θ_2 through n increments of this angle.

The progression of a slider-crank mechanism program calculates the displacement, velocity and acceleration, the maximum and minimum displacements, and variation of the angular value for the connecting rod angle θ_3. Data such as the crankshaft speed in rpm, the slider offset, and other dimensions must be stored with the angular velocity of the crank in order to run the program. Again, it uses basic equations of motion (see any textbook on kinematics) which when differentiated yield the required motion characteristics.

Finally, the program used in determining the required moments of inertia as shown in Figures 2-24 and 2-26 was also quite helpful.

Appendix 2

Solutions to Selected Review Problems, Chapter 2

Problem 1

Given:

$T_2 = 150$ in.-lb

Find:

F_4 to balance the given torque

Use:

$K_S = 1{:}10$

$K_F = 1{:}40$

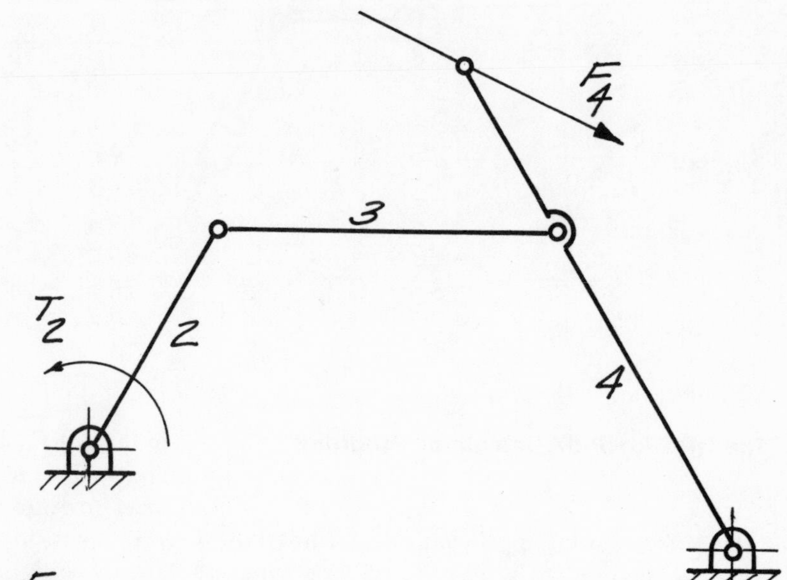

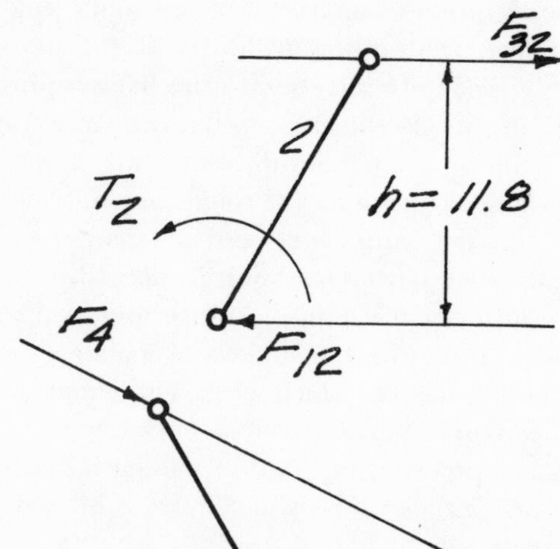

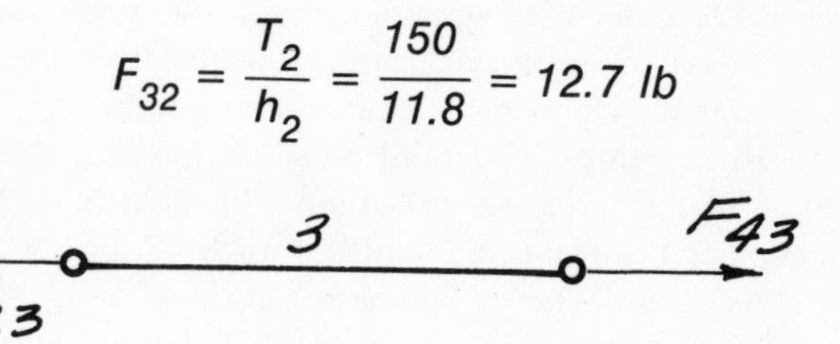

$$F_{32} = \frac{T_2}{h_2} = \frac{150}{11.8} = 12.7 \text{ lb}$$

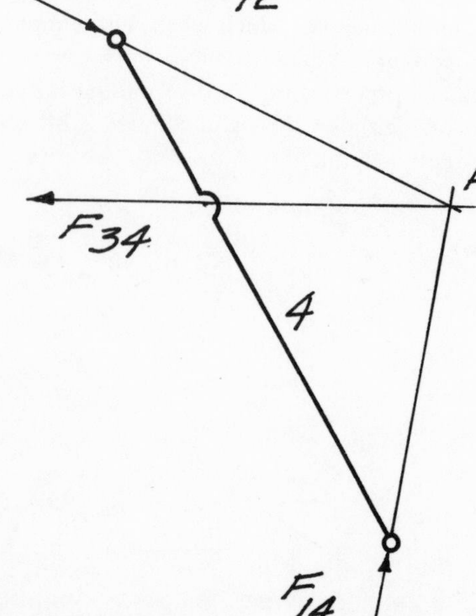

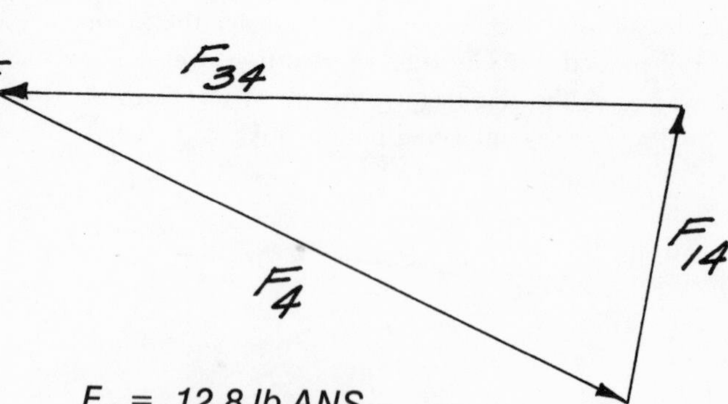

$F_4 =$ 12.8 lb ANS.
(Measured from Polygon)

Problem 2

Given:

$F_6 = 1$ kip

Find:

F_{cylinder}

Use:

$K_S = 1{:}10$

$K_F = 1{:}200$ lb

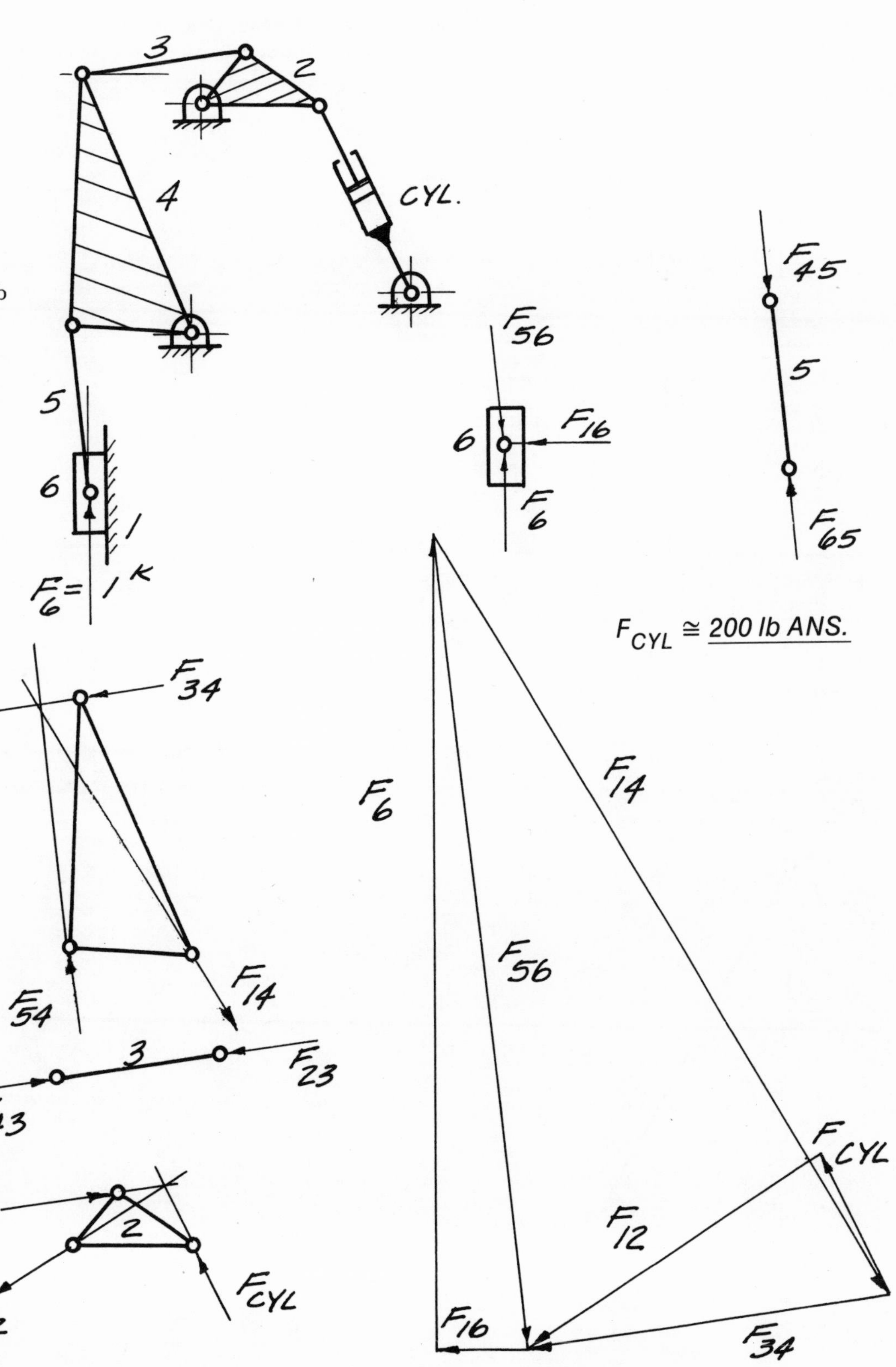

Problem 4

Derivation of equation:

$$\frac{b}{\sin \beta} = \frac{c}{\sin \theta_2}$$

$$\therefore \sin \beta = \frac{b \sin \theta_2}{c}$$

$$T_2 = \frac{F b \sin(\theta_2 + \beta)}{\cos \beta} = \frac{F b (\sin \theta_2 \cos \beta + \sin \beta \cos \theta_2)}{\cos \beta}$$

$$= F b \sin \theta_2 + \frac{F b^2 \sin \theta_2 \cos \theta_2}{c \cos \beta};$$

substituting

$$\cos^2 \beta = 1 - \sin^2 \beta = 1 - \frac{b^2}{c^2} \sin^2 \theta_2$$

yields

$$T_2 = F b \sin \theta_2 \left[1 + \frac{b \cos \theta_2}{c\sqrt{1 - \left(\frac{b}{c}\right)^2 \sin^2 \theta_2}} \right];$$

The TI-59 pocket calculator program results are shown.

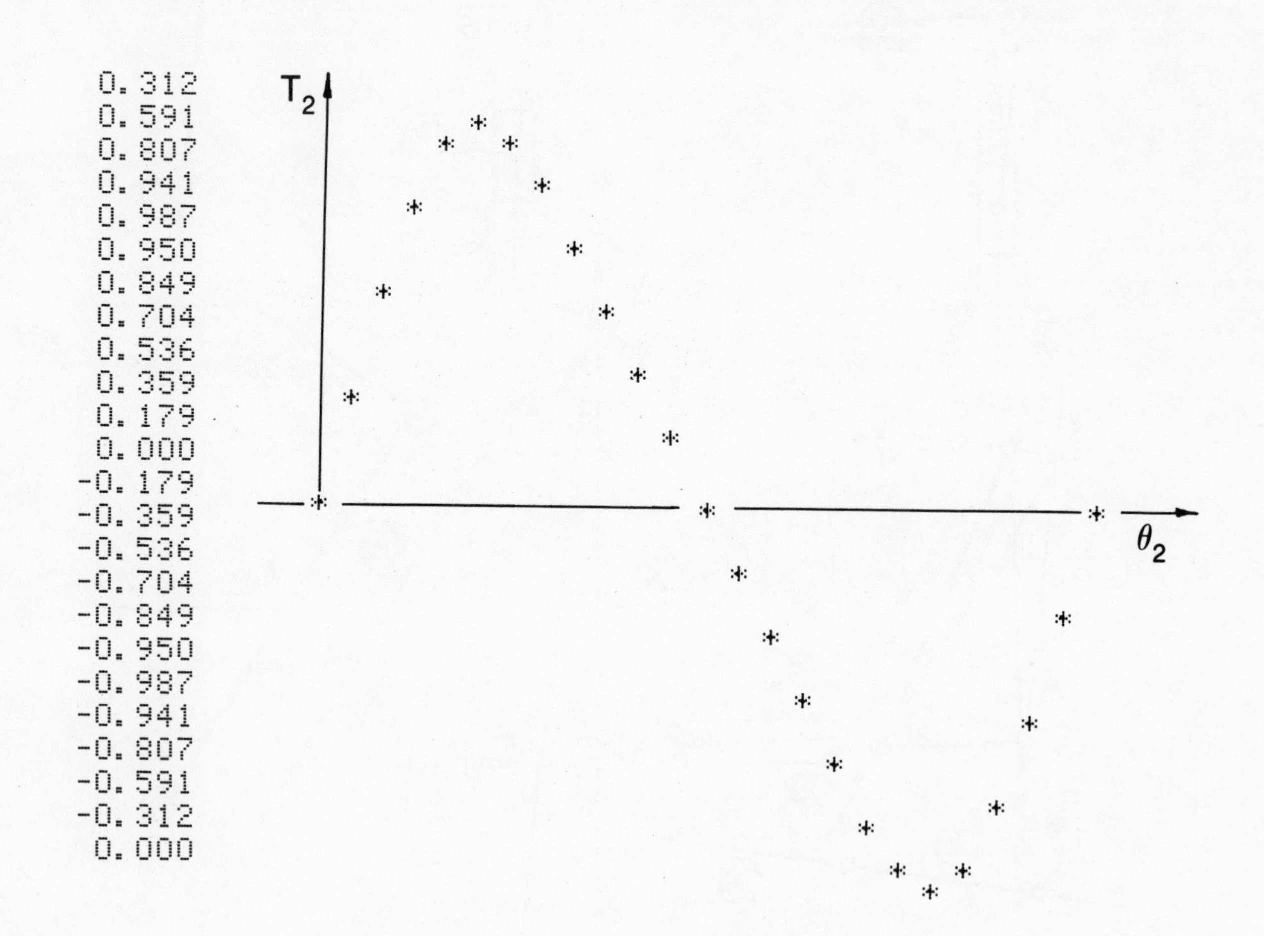

Problem 9

Given:

$\omega_2 = 200$ rad/sec ccw, const.
$\theta_2 = 150$ deg
$P_1 = 50$ lb
$P_2 = 70$ lb
$O_2A = 2$ in.
$AB = 6$ in.
$BO_4 = 5$ in.
$O_2O_4 = 6$ in.
$W_3 = 0.7$ lb
$W_4 = 0.8$ lb
$r_{G3} = 4$ in.
$I_3 = 0.02$ lb in.-sec²
$I_4 = 0.01$ lb in.-sec²
$r_{G4} = 2.5$ in.

Find: (Graphically and Analytically)

$\omega_3 = 30.97$ rad/sec
$\omega_4 = 70.97$ rad/sec
$A_{G3}\vec{i} = 49{,}676$ in./sec²
$A_{G4}\vec{i} =$ ______ in./sec²
$h_3 = 1.61$ in.
$\alpha_3 = 7{,}612$ rad/sec²
$\alpha_4 = 6{,}250$ rad/sec²
$A_{G3}\vec{j} = 16{,}387$ in. /sec²; $\theta_A = 18.26°$
$A_{G4}\vec{j} =$ ______ in/sec²; $\theta_{AG4} =$ ______°
$h_4 =$ ______ in.

Results:

$F^{I}_{23,\ m_3}$	$F^{II}_{23,\ m_4}$	$F^{III}_{23,\ P_1}$	$F^{IV}_{23,\ P_2}$	ΣF_{23}	T_2
lb	lb	lb	lb	lb	in.-lb
79		43		108	140

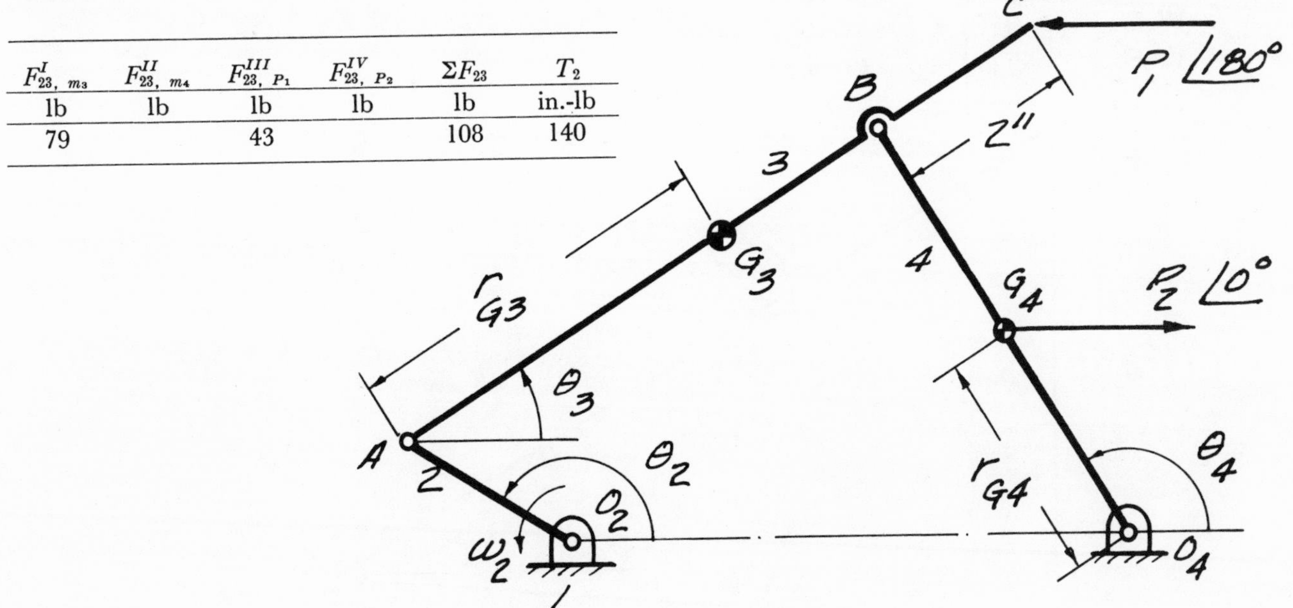

Solutions to Selected Review Problems, Chapter 3

Problem 1

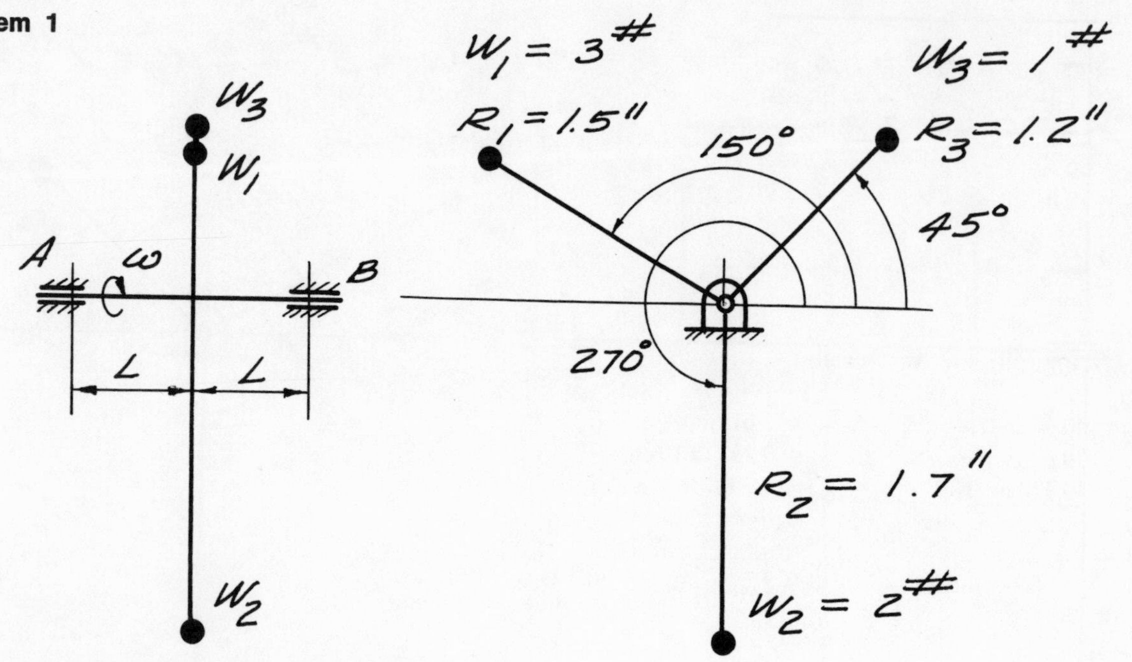

$W_1R_1 = (1.5)(3) = 4.5$ in.-lb
$W_2R_2 = (1.7)(2) = 3.4$ in.-lb
$W_3R_3 = (1.2)(1) = 1.2$ in.-lb

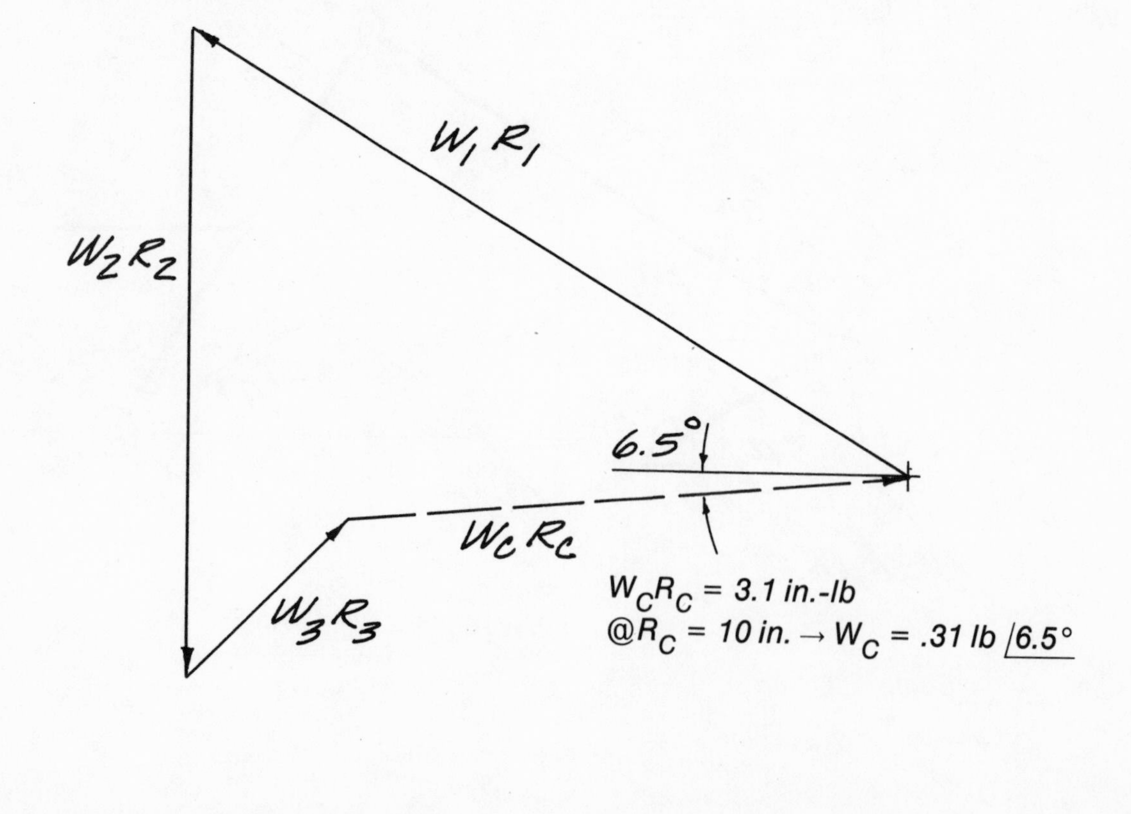

$W_CR_C = 3.1$ in.-lb
@$R_C = 10$ in. → $W_C = .31$ lb $\angle 6.5°$

Problem 2

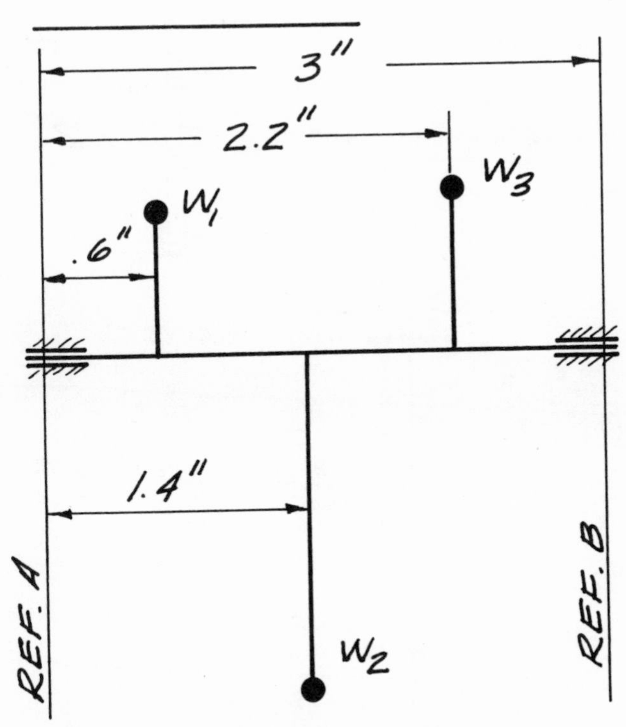

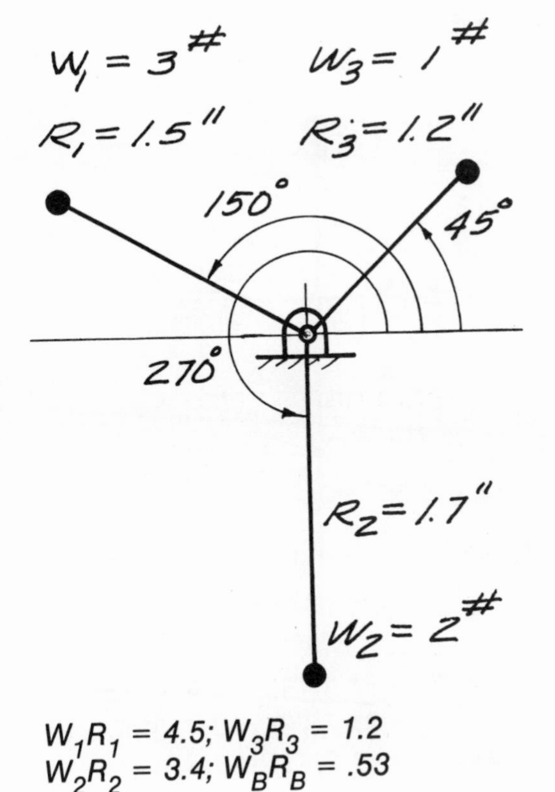

$W_1R_1X_1 = (3)(1.5)(.6) = 2.7$
$W_2R_2X_2 = (2)(1.7)(1.4) = 4.8$
$W_3R_3X_3 = (1)(1.2)(2.2) = 2.6$

$W_1R_1 = 4.5$; $W_3R_3 = 1.2$
$W_2R_2 = 3.4$; $W_BR_B = .53$

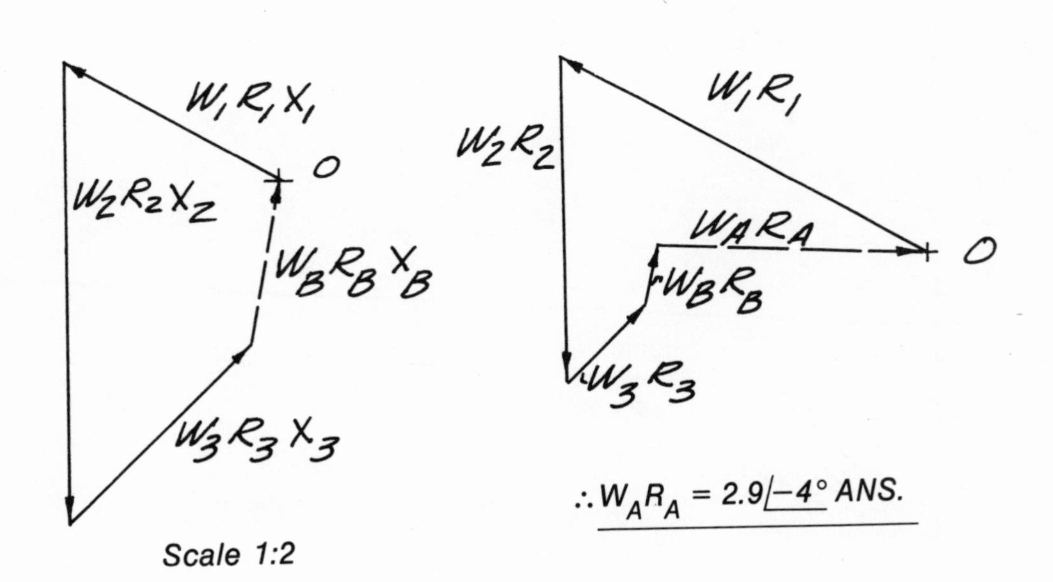

Scale 1:2

$\therefore W_AR_A = 2.9\angle{-4°}$ ANS.

Analytical Solution for Problem 2

Given: Data same as graphical solution to Problem 2, and $\omega = 100$ rad/sec

Find: Balance machine

Solution:

$$M_1R_1\omega^2 = \left(\frac{4.5}{32.2}\right)\left(\frac{100^2}{12}\right) = 116.46 \text{ lb } \underline{/150^\circ}$$

$$= -100.86\vec{k} + 58.23\vec{j}$$

$$M_2R_2\omega^2 = \left(\frac{3.4}{32.2}\right)\left(\frac{100^2}{12}\right) = -87.99\vec{j}$$

$$M_3R_3\omega^2 = \left(\frac{1.2}{32.2}\right)\left(\frac{100^2}{12}\right) = 31.06 \text{ lb } \underline{/45^\circ}$$

$$= 21.96\vec{k} + 21.96\vec{j}$$

$$\Sigma\vec{M}_A = 0; (.6\vec{i}) \times (-100.86\vec{k} + 58.23\vec{j}) + (1.4\vec{i}) \times (-87.99\vec{j}) + (2.2\vec{i}) \times (21.96\vec{j} + 21.96\vec{k}) + (3\vec{i}) \times (\vec{F}_B) = 0$$

Hence,

$$12.21\vec{j} - 39.94\vec{k} + 3F_B\vec{k} - 3F_{Bj}\vec{} = 0$$

and

$$\vec{F}_B = 4.07\vec{k} + 13.31\vec{j} = 13.92 \text{ lb } \underline{/73^\circ}$$

$$\Sigma\vec{M}_B = 0; (-.8\vec{i}) \times (21.96\vec{j} + 21.96\vec{k}) + (-1.6i) \times (-87.99\vec{j}) + (-2.4\vec{i}) \times (-100.86\vec{k} + 58.23\vec{j}) + (-3\vec{i}) \times (\vec{F}_A) = 0$$

and

$$\vec{F}_A = 74.83\vec{k} - 5.51\vec{j} = 75.03 \underline{/-4.2^\circ}$$

then

$$F = Ma = \frac{W}{g} R_A\omega^2$$

or

$$W_AR_A = \frac{386\, F_A}{\omega^2} = 2.90 \underline{/-4.2^\circ} \text{ ANS.}$$

and

$$W_BR_B = \frac{386}{\omega^2} F_B = 0.54 \underline{/73^\circ} \text{ ANS.}$$

Index